Jeff Daniel Nze Memiaghe

Agricultural Practices and Spatial Variability of Phosphorus in Soils

Jeff Daniel Nze Memiaghe

Agricultural Practices and Spatial Variability of Phosphorus in Soils

ScienciaScripts

Imprint

Any brand names and product names mentioned in this book are subject to trademark, brand or patent protection and are trademarks or registered trademarks of their respective holders. The use of brand names, product names, common names, trade names, product descriptions etc. even without a particular marking in this work is in no way to be construed to mean that such names may be regarded as unrestricted in respect of trademark and brand protection legislation and could thus be used by anyone.

Cover image: www.ingimage.com

This book is a translation from the original published under ISBN 978-620-6-70890-2.

Publisher:
Sciencia Scripts
is a trademark of
Dodo Books Indian Ocean Ltd. and OmniScriptum S.R.L publishing group

120 High Road, East Finchley, London, N2 9ED, United Kingdom
Str. Armeneasca 28/1, office 1, Chisinau MD-2012, Republic of Moldova, Europe
Printed at: see last page
ISBN: 978-620-8-07414-2

Contents

Summary

Phosphorus (P) is an important nutrient for plant growth. However, excessive applications of P to agricultural soils can increase the risk of loss of this element through surface runoff and contribute to eutrophication of watercourses. Sustainable management of P in arable soils therefore depends on fertilisation that balances soil P availability with the actual P demand of plants. It is therefore important to have a better understanding of the spatial variability (SV) of P available to crops, in order to improve the economic and rational use of phosphate fertilisers, promote the profitability and sustainability of the farming business, while reducing P losses to the environment. The general objective of this PhD thesis was to evaluate the SV of soil P availability under different cropping systems (old grassland (OG) vs young grassland (YP)) and cropping practices (conventional tillage (CT) vs direct seeding (DS)) using statistical and geostatistical tools for agri-environmental recommendations.

To meet the objectives of the thesis, four commercial crop fields were used, two in the Chaudière-Appalaches region and two in Montérégie (province of Quebec). Two soils from the Chaudière-Appalaches region, one under young grassland (JP; 2.4 ha; 2 years) and one under old grassland (AP; 2.5 ha; 10 years under permanent pasture), classified as Humo-Ferric Podzols, received organic amendments (manure, cattle slurry). The results revealed an average P saturation index (PSI, $(P/A1)_{M3}$) of 3% in both layers (0-5 cm and 5-20 cm) under JP compared with 7% in the 0-5 cm layer and 4% in the 5-20 cm layer under AP, following long-term applications of manure. For both fields under grassland (JP and AP), we observed a high intensity of P variability (CV > 50%) and a moderate spatial structure (25-75%). Repeated applications of cattle manure and slurry may have a long-term impact on soil P accumulation, reducing P variability and spatial dependence in permanent grassland. Due to the observed stratification of P, a soil sampling strategy focusing on the 0-5 cm layer should be retained in permanent grasslands (PS) for sustainable P fertilizer recommendations in soils of the Province of Quebec.

In addition, this study examined the impacts of tillage on the spatial variability of available P in two commercial fields that have been in corn-soybean rotation for 20 years, labelled TC (10.8 ha) and SD (9.5 ha), for the purpose of improving phosphate fertilizer recommendations in the Montérégie region. The results showed that the variability of available soil P in the two fields ranged from moderate to very high (32-60%) under TC and SD. An average PSI of 3% was measured in both layers under CT, compared with 8% in the 0-5 cm layer and 6% in the 5-20 cm layer under SD. Relationships between P indices and other properties physico-chemical properties (CT, Fe_{M3}, pH_{water}, Ca_{M3}) differed according to tillage practices. This study demonstrated the importance of variable rate application (VRA) for improving P fertilizer recommendations in the Province of Quebec.

In addition, the delimitation of management zones (ZAs) was carried out in the field under SD (9.5 ha). The ZAs were delimited on the basis of the apparent electrical conductivity (ECa)-soil P relationship in order to reduce the spatial variability of soil P for field-specific P fertiliser recommendations. The ECa data (ECa_{30}; ECa_{100}) were measured at two depths (0-30; 0-100 cm) using the Veris. ZAs were delimited using the ISODATA method according to three delimitation strategies: PSI measurements, ECa measurements, or combined PSI+ECa measurements. The results showed that the average PSI was 7.9% under SD. Soil P variability was moderate (32-36%), indicating that a uniform P fertilizer recommendation cannot be applied in the field. Mean ECa_{30} and ECa_{100} values were 15.8 and 32.6 mS m^{-1} respectively. A significant correlation between ECa_{30} and soil P (0.22-0.23) was observed. Delineation into two-three ZAs using PSI measurements generated the greatest reductions in P (40-74 kg P_2O_5), thus representing the best agronomic strategy for delimiting ZAs. This study highlighted the potential for reducing the spatial variability of soil P using the ZAs delimitation strategy, thereby reducing P losses under SD.

Thus, the ATV generated 366 kg P_2O_5, while the two-three ZAs delimitation generated 306 and 341 kg P2O5 in the field under SD, i.e. gains of 60 and 25 kg P2O5 respectively compared to the ATV in the grain maize crop. Consequently, the two-three ZAs delimitation strategy appeared to be more effective than the ATV in terms of P2O5 gains in this field. Overall, the results of this thesis have demonstrated the importance of taking into account the SV of P for specific P management purposes, in order to reduce P losses in soils under field crops in precision agriculture.

Abstract

Phosphorus (P) represents an important soil nutrient for plant growth. However, excessive P applications in agricultural soils relative to crop requirements may increase P loss risk through runoff, contributing to water eutrophication. Thus, field-scale sustainable management of soil P should be based on balanced fertilization between soil P supply and P real demand from crop requirements. Therefore, it is important to have a better understanding of spatial variability of soil available P under intensive managed crops, aiming to improve economic and rational use of P fertilizers, to promote profitability and sustainability of agricultural companies, while reducing P losses in the environment. The main goal of this PhD dissertation was to evaluate spatial variability of soil available P under contrasted crops [old grassland (OG) vs young grassland (YG)] and tillage systems [conventional (CT) vs no-tillage (NT)] using statistical and geostatistical tools, aiming to provide agri-environmental recommendations.

To reach the goals of this PhD dissertation, four commercial fields were used. Two fields were located in Chaudière-Appalaches region and two others in Montérégie region (Eastern Canada). Thus, the study aimed to investigate the field-scale variability of soil-test P (STP) under two contrasting grassland fields in Chaudière-Appalaches region using descriptive statistics and geostatistics for accurate recommendations on soil sampling strategy and sustainable approaches to P management. A young grassland (YG; 2 years) and an old grassland (OG; 10 years under permanent pasture) were classified as humo-ferric podzol and received organic fertilizers. Soil samples were collected in 16-m by 16-m triangular grids at two depths (0-5 and 5-20 cm). Results showed a P saturation index (PSI, $(P/A1)_{M3}$) mean value of 3% in both YG soil layers (0-5 cm and 5-20 cm) against 7% in OG 0-5 cm layer and 4% in OG layer 5-20 cm, owing to long-term manure applications. Variability of soil P was high (CV > 50%) and spatial structure of soil P was moderate (25-75%) under both contrasted fields. Repeated applications of cattle and slurry manure may have a long-term impact on soil P accumulation, thereby reducing spatial variability and dependence of soil P under permanent grasslands. Owing to P stratification, a soil sampling strategy focused on 0-5 cm layer should be retained under permanent grasslands (OG) for sustainable P fertilizer recommendations in Eastern Canada.

Moreover, the study examined tillage impacts on spatial variability of soil-available P, in two commercial fields under 20 years maize-soybean rotation, denoted CT (10.8 ha) and NT (9.5 ha), aiming to improve P fertilizer recommendations in Eastern Canada. Blends of NPK fertilizers were applied to the soils (Humic Gleysols) following local recommendations. Results showed that the variability of soil available P ranged from moderate to very high (32-60%) under CT and NT. The mean PSI value was 3% under both CT layers, against 8% and 6% under the 0-5 cm and 5-20 cm NT layers, respectively. Relationships between P indices and other soil chemical properties (TC, Fe_{M3}, pH_{water}, and Ca_{M3}) differed between contrasted tillage practices. This study highlighted the importance of variable rate application (VRA) to improve P fertilizer recommendations in Eastern Canada.

Furthermore, delineating management zones (MZs) was performed within the 9.5 ha NT field. This MZ delineation approach was based on apparent electrical conductivity (ECa)-soil P relationship, aiming to reduce spatial variability of soil P for predicting field-specific P fertilizer recommendations. The ECa data (ECa_{30}; ECa_{100}) were measured at two soil depths (0-30; 0-100 cm) using Veris system. The MZs were delineated using ISODATA method following three delineation strategies: ISP, ECa, and combined ISP+ECa measurements. Results showed that mean ISP value was 7.9% under NT field. Variability of soil P was moderate (32-36%), indicating that uniform P fertilizer recommendation cannot be applied in the field. The ECa_{30} and ECa_{100} mean values were 15.8 and 32.6 mS m^{-1} , respectively. A significant correlation between ECa_{30} and soil P (0.22-0.23) was observed. Delineating the field into two-three MZs using ISP measurements generated the highest P reductions (40-74 kg P_2O_5), thus representing the best agronomic strategy for delineating MZs. This study highlighted the importance of using MZ delineation strategy to reduce the spatial variability of soil P, decreasing P fertilizer applications and P losses under the NT field.

Thus, VRA agronomy strategy generated 366 kg P2O5 , while delineation strategy into two-three MZs generated 306 and 341 kg P_2O_5, respectively under the NT field. This represented 60 and 25 kg P2O5 reductions compared to the VRA under corn-grain NT field. Therefore, delineation strategy into two-three MZs seemed more effective compared to VRA strategy in terms of P2O5 reduction in this field-scale. The results of this PhD dissertation demonstrated the importance for taking into account the spatial variability of soil P for field-specific P management, aiming to reduce P losses under large crop fields in precision agriculture.

To my parents and role models. Dr Ing. Joseph-Aimé Memiaghe and Dr. Gisèle Memiaghe née Ossakedjombo-Ngoua. Gisèle Memiaghe née Ossakedjombo-Ngoua. Thank you for the most precious gift you have given me: Life.

Acknowledgements

This thesis marks the definitive end of my university studies. In fact, it is the culmination of a life project that began when I arrived in Canada, at the start of my B.Sc. programme in Agronomy, with a major in Soils and the Environment, at the Faculty of Agricultural and Food Sciences of ¡1 1 Université Laval in 2004. Throughout my academic career, many people and professors have supported and encouraged me. It is difficult for me to mention them all by name. However, I would particularly like to thank :

My thesis co-supervisor, Dr Athyna N. Cambouris, a researcher at Agriculture and Agri-Food Canada's (AAFC) Quebec Research and Development Centre, for her guidance and all the facilities and opportunities she gave me. Her scientific and moral support, her calmness and patience helped me to improve day by day. Thank you Athyna, for opening the doors to applied geostatistics in Precision Agriculture!

My thesis supervisor, Dr Antoine Karam, professor at Laval University, for his availability, his encouragement at the various conferences we attended together, and his scientific rigour. Antoine, I would like to express my deepest gratitude for the facilities and opportunities you have given me. Thank you for welcoming me into your team, which has since become my second family!

Dr Noura Ziadi for her essential and enriching help in writing the chapters of this thesis. Thank you, Noura, for your critical spirit and scientific rigour.

I would also like to thank the external assessors of this thesis, Dr Aimé Jean Messiga, Dr Judith Nyraneza and Dr Louis Longchamps. Through your comments and suggestions, you have greatly contributed to the improvement of this thesis.

I would like to thank Marc Duchemin for his technical support in producing the many figures and tables in the chapters of this thesis. Marc, thank you for your suggestions and criticisms, because without your observations, this work might not have seen the light of day. Thank you very much!

I'd like to thank my friends from AAC: Chedzer Clarc-Clément, Wen Guoqi, Éric Manirakiza, Karim Lajili, Nomena Ravelojaona, and Bilal Javed for all the discussions, laughs and good times we've had together. I'd also like to extend my warmest thanks to my friends (brothers): Piterson Floradin, Raghad Soufan, Dacos (Wally) Ndiaye, Ralph-Dimitri Tasing, Thierno Diallo, Pierre-Paul Audate, Pôl Zué Essangui, Patrick Kankolongo, Siaka Koné, and my dear aunt (mother) Zedna-Inès Étsang- Metegue épouse Yèno for advising and supporting me during my doctoral studies.

This thesis was made possible thanks to the financial support of AAFC. I would also like to thank the Department of Soils and Agri-Food Engineering at Laval University and Centre SÈVE for awarding me grants to present my research results at scientific conferences. In addition, the results of this doctoral thesis have been the subject of several prestigious awards, including the *Karl Ivarson Soil Science* Prize ($3,000) offered by the Canadian Agri-Food Education Foundation in 2023, the *Yvon Lévesque* Prize ($400) offered by the Ordre des Agronomes du Québec in 2023, the *ISPA Outstanding Graduate Student Award* ($1,000) from the International Society for Precision Agriculture in the United States (2022-2024), the International Union of Soil Science scholarship (£300) as part of the World Soil Science Congress 2022 and a doctoral scholarship ($5,000) offered by Université Laval in 2022.

I dedicate this thesis to my late paternal and maternal parents. My paternal uncles (fathers) left too soon: Dr Antoine-Joel Olame-Nze, Gabriel Nze-Bithegue and Jean-Christian Nkouele Nze-Memiaghe. You have been and continue to be true role models for me. To my dearest maternal grandmother, Georgette Ntombwiyandji Tétaye Simira, widow of Martin Ngoua, thank you for your love, your gentleness, your humility, your silence, your patience and your wise advice to your grandchildren! May you be with your Father, whom you taught us to honour from an early age! To my maternal great-uncles: Dr Albert Ndjavé-Ndjoy, a military doctor, and Pascal Ndjavé Émane, thank you for your tireless devotion to the family! May God repay you a hundredfold!

Finally, I would like to thank my parents: Dr. Ing. Joseph-Aimé Memiaghe and Dr. Gisèle Marie- Hortense Memiaghe, née Ossakedjombo-Ngoua, my brothers and sisters: Nanou, Ange, Clisy-Awan, and Ludmillia for their support and encouragement throughout my many years of study. Thank you Dad for your suggestions and advice during our scientific exchanges. Thank you too, Mum, for your advice, the values you have passed on and, above all, the rigour you have shown us since our earliest childhood, inherited from your dear parents.

Foreword

This doctoral thesis is part of a collaboration between ¡Laval University and Agriculture and Agri-Food Canada's Quebec Research and Development Centre.

The document is divided into five chapters. The first chapter describes the scientific literature review, in which topics related to the subject of study are presented, such as the phosphorus cycle and methods for determining soil phosphorus, agronomic and environmental factors affecting phosphorus stratification, the current state of knowledge on the spatial variability of phosphorus in soils, and the main agronomic strategies for controlling the spatial variability of phosphorus in soils under field crops.

Chapters 2, 3 and 4 are scientific articles written in English. Chapter 2 deals with the spatial variability of phosphorus indicators under two contrasting grassland systems. This article was published in December 2020 in *Agronomy MDPI Journal*. Chapter 3 is an article published in May 2022 in *Soil Systems MDPI Journal*. It deals with the impacts of tillage on the spatial variability of phosphorus under maize-soybean rotation. Chapter 4 presents management zone delimitation strategies to reduce the spatial variability of soil phosphorus for phosphate fertilizer recommendations in a no-till field. This article will be submitted to the *Canadian Journal of Soil Science*. The articles were co-authored with Dr. Athyna N. Cambouris, Dr. Noura Ziadi, Dr. Marc Duchemin, Isabelle Perron of the Agriculture and Agri-Food Canada Research and Development Centre in Quebec City, and Dr. Antoine Karam of Université Laval. All articles have been included in the thesis as submitted. Chapter 5 includes a general discussion related to the previous chapters as well as a conclusion and future recommendations.

Field experiments, data collection and soil and plant analysis in the laboratory were carried out by Dr. Athyna Cambouris' team of research professionals. Athyna Cambouris' team of research professionals. Data processing and analysis were carried out by myself under the supervision of Dr. Athyna Cambouris. Athyna Cambouris.

The results of these studies were presented at the International Congress of the *American Society of Soil Scientists* in San Antonio, USA, in November 2019, at the 35ème congress of the *Association Québécoise des Spécialistes en Science du Sol* in Quebec City, in June 2021, at the 15ème *International Conference on Precision Agriculture* in Minneapolis, USA in June 2022, the 22ème *World Congress of Soil Science* (in virtual mode) in Glasgow, UK in August 2022, the International Congress of the *American Society of Soil Scientists* in Baltimore, USA in November 2022, the scientific meetings of the Ordre des Agronomes du Québec in Lévis in April 2023 as part of the presentation of the *Yvon Levesque* prize, and the International Water Congress in Quebec City in May 2023.

Introduction

Phosphorus (P) accounts for 0.12% of the lithosphere's constituent elements, making it the 11ème most abundant element in the Earth's crust (Cathcart, 1980; Messiga, 2010; Li, 2017). The phosphate content of phosphate-bearing rocks from nature reserves is between 100 and 140 g P kg^{-1} . These natural reserves are the main sources of phosphates in the world (Morel, 2002). Recently, the world's phosphate reserves were re-evaluated at 300 billion metric tonnes, due to new phosphate deposits discovered in North Africa, China, the Middle East and the United States (US Geological Survey, 2022).

The distribution of the world's phosphate reserves and the availability of mineable deposits vary widely depending on their geographical distribution. The main natural phosphate deposits are located in six countries (Morocco (70%), China (5%), Egypt (4%), Algeria (3%), South Africa (2%) and Brazil (2%)) (US Geological Survey, 2022). Due to the high global demand for phosphates in the face of a growing world population, it is generally recognised that phosphate reserves are diminishing considerably and the current practice of phosphate extraction is no longer in line with the sustainable development advocated (Steen, 1998; Gassner, 2003; Fu et al., 2013a).

P is an important nutrient for the growth and development of cultivated plants. It is used in the formation of nucleic acids and plays a key role in crop growth (Fu et al., 2013a). Excessive application of P to agricultural soils in relation to crop needs leads to an accumulation of P in the topsoil of agricultural plots, increasing the risk of loss of this element through surface runoff (Jamieson et al., 2003) via erosion of soil particles. This contributes to environmental degradation through eutrophication of watercourses (Grant et al., 1996; Jordan et al., 2000; Messiga, 2010; Wei-Xia et al., 2015).

Thus, sustainable P management in arable soils relies on fertilisation that balances the actual soil P supply and demand (Tunney, 1990; Fu et al., 2013a). It is therefore important to have a better understanding of the spatial variability of P in crop soils, in order to improve the economic and rational use of P fertilisers, promote the profitability and sustainability of the farming enterprise, improve the reduction of P losses and environmental protection.

Several studies have been carried out on the spatial variability of P at the catchment scale (Nolan et al., 2007; Yuan et al., 2009; Wilson et al., 2016), or at the regional and territorial scale (Roger et al., 2014; Wei-Xia et al., 2015; Vasu et al., 2017). Most of these studies focus on planning and mapping available P in soil for the purposes of prevention and assessment of the risk of P-related environmental pollution. However, very few studies have been conducted on the spatial variability of available P at the scale of an agricultural plot under the influence of different agronomic practices (cropping systems, tillage) in the province of Quebec. In addition, to date there have been few studies of management zones based on monitoring P variability for the purposes of field-scale P recommendations.

The general objective of this PhD thesis is to assess the spatial variability of soil phosphorus availability under different cropping systems and tillage practices for the purposes of agri-environmental P recommendations in the context of precision agriculture.

Chapter 1

Literature review

1.1. Phosphorus: biogeochemical cycle, chemical forms and methods for determining available P in cultivated ecosystems

1.1.1. Biogeochemical cycle of P in cultivated ecosystems

The main sources of P in cultivated soils are mineral fertilisers, organic farm amendments (manure, slurry), fertilising residual matter (sewage sludge, compost) and atmospheric deposition (Morel, 2007; Gagnon et al., 2010) (Figure 1.1). In Quebec, Hébert et al. (2008) report that organic farm amendments (manure and slurry) account for 63% of P inputs to agriculture, while mineral fertilizers and fertilizer residues account for 35% and 2% of P inputs respectively. More than 31 million tonnes of organic amendments were produced by Quebec farms for spreading on agricultural fields, representing more than 95,000 t of P2O5 (MDDELCC, 2017). Crop residues are also a major source of P in cultivated soils.

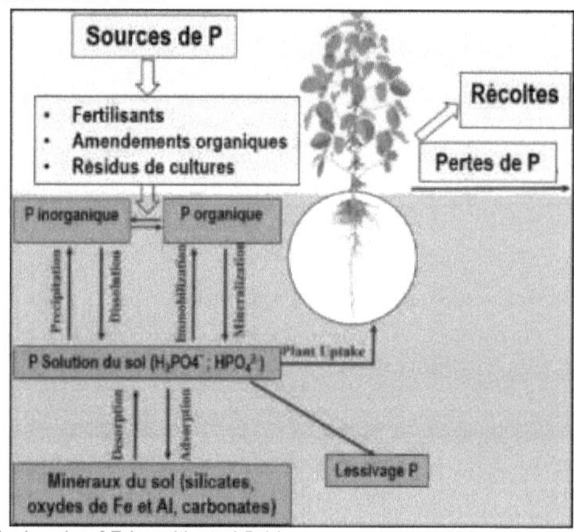

Figure 1-1. Biogeochemical cycle of P in cultivated fields

(Source: adapted from Fageria et al. (2017))

The different pools are inorganic P, organic P and mineral particles (silicates, carbonates, Fe and Al oxides). They interact with the soil solution through physicochemical, biological and biochemical mechanisms. These mechanisms are mainly adsorption, desorption, precipitation, dissolution, mineralisation, immobilisation, intra-particle diffusion, oxidation-reduction and acid-base reactions. All these reactions can lead to the release of P ions into the soil solution or to the subtraction of P ions from the soil solution into the solid phase.

Part of the P is exported outside the cultivated plot via harvested products and aerial plant parts. For example, Webb et al (1992) showed that average annual P exports were respectively 60 kg P2O5 ha-1 for maize and 32 kg P2O5 for soybeans, i.e. an average of 46 kg P2O5 ha-1 in fine loamy calcareous soils in Iowa, resulting from 14 years of phosphate fertilisation under a soybean-maize rotation. In Quebec, Giroux et al (2002) also reported that an average of 2.5 and 3 kg P2O5 t-1 were removed from grass and legume meadows respectively.

Other P outputs are also assessed in the form of loss by leaching (deep migration of dissolved P), leaching (deep migration of particulate P), surface runoff and preferential flow (Simard et al., 2000; Djodjic et al., 2004; Allaire et al., 2011; Abdi et al., 2014). In addition, new knowledge on P losses in surface waters and lakes requires better quantification of P losses (Simard et al., 2000; Giroux et al., 2008; Allaire et al., 2011). In their report on changes in soil P levels according to their fertility, Giroux et al. (2002) state that the effects of plant P exports on changes in soil P cannot be properly measured without taking account of the often fairly high spatial variability of soil P levels, hence the need for a better understanding of the spatial variability of soil P.

1.1.2. Chemical forms of P in soils

P is present in cultivated soils in 3 forms: inorganic P, organic P and P linked to microbial biomass, which represent on average 75%, 20% and 5% of total P, respectively (Grant et al., 2005).

Inorganic P in the soil is distributed between (1) the soil solution in the form of the dominant orthophosphate ions ($H_2PO_4^-$ and HPO_4^{2-}), (2) phosphates adsorbed on the solid phase, and (3) phosphates precipitated with iron and aluminium oxides and hydroxides, and calcium carbonates (Pierzynski et al., 2005). In acid soils, phosphates are bound to iron and aluminium oxides to form minerals such as strengite ($FePO_4.2H_2O$) and variscite ($AlPO_4.2H_2O$), while in neutral and alkaline soils, phosphate compounds are bound to calcium to form minerals such as calcium phosphates, magnesium phosphates and the various forms of apatite ($Ca_3(PO_4)_2CaF_2$, $Ca_3(PO_4)_2CaF_2$, $Ca_3(PO_4)_2CaCl_2$, $Ca_5(PO_4)_3OH$, $Ca_{10}(PO_4)_6(CO_3)$). In this way, P is bound to iron and aluminium under acidic conditions and to calcium under neutral and alkaline conditions (Pierzynski et al., 2005).

Figure 1.2 summarises the effect of soil solution pH on the distribution of orthophosphate ions in solution. The dominant ionic forms of P ($H_2PO_4^-$ and HPO_4^{2-}) are influenced by the pH of the soil solution. At a pH below 7.2, the $H_2PO_4^-$ form is dominant in the soil solution, while at a pH above 7.2, the HPO_4^{2-} form predominates in the soil solution.

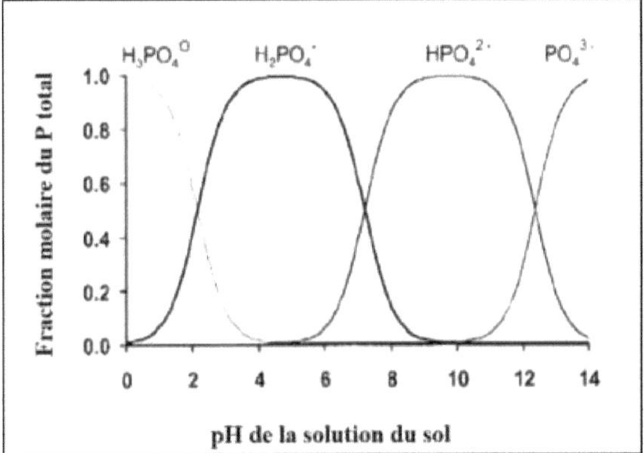

Figure 1-2. Effect of soil solution pH on the distribution of orthophosphate ions in solution

(Source: adapted from Hinsinger, 2001)

Furthermore, Abdi et al (2014) identified, using the nuclear magnetic resonance (NMR) technique, three main forms of soil inorganic P (orthophosphates, pyrophosphates and polyphosphates) in their study of the long-term impact of cropping practices (conventional tillage vs no-till) and phosphate fertilisation on soil P forms. The researchers found that the orthophosphate form is the most dominant (over 95%), followed by the pyrophosphate and polyphosphate forms. In fact, the orthophosphate group represented 95.94% and 96.5% of inorganic P forms in conventionally tilled and no-till soils respectively. The proportion of orthophosphate ions could increase with soil depth. Indeed, Abdi et al. (2014) observed that orthophosphate ion fractions represented 95.65%, 96.49% and 96.16% of total inorganic P in soil samples from the 0-5 cm, 5-10 cm, and 10-20 cm horizons, respectively. Thus, the proportion of orthophosphate ions could vary according to the type of cultivation practice and soil depth. When soils receive P in the form of soil improvers, the concentration of phosphate ions in the solution can increase depending on the phosphate status and soil type (Kovar and Barber, 1988; Messiga, 2010). This increase in phosphate ion concentration is beneficial for crop growth, but could also pose a risk to the environment.

Soil organic P is defined as P linked to carbon. It comes from animal and plant residues (Kuo, 1996). The organic fraction of P represents around 20% of the total P in the topsoil of soils and could be as much as 50% or even 80% of the total P (Stevenson 1994, Roger et al., 2014). Its content in soils depends on that of organic matter, which contains 0.5% P. The release of organic P into the soil solution is controlled by the mineralisation of organic matter.

Soil organic P can be grouped into four groups, namely the orthophosphate monoester group, which are phosphoric acid esters including inositol phosphates, phosphate sugars, phosphoproteins and mononucleotides. The diester phosphate group, which includes DNA, RNA, phospholipids, teichoic acid and aromatic compounds. The phosphonate group, which includes phosphonic acids and phosphonolipids characterised by C-P bonds instead of ester bonds. Finally, there is the group of anhydrous orthophosphates (Smith and Read, 1997; Corbridge, 2000; O'Halloran and Cade-Menum, 2007). However, Turner et al (2005) classify soil organic P into three groups: phosphate esters, phosphonates and phosphoric acid anhydrides. This classification is essentially based on the chemical nature of the P bonds.

Abdi et al, (2014) identified three classes of soil organic P including (1) phosphonates, (2) orthophosphate monoesters and (3) orthophosphate diesters in their study on the long-term impact of tillage (conventional vs. no-till) and phosphate fertilisation on soil P forms. Moreover, the phosphonate group is stable and not very bioavailable in soils (Condron et al., 2005). Figure 1.3 shows the various organic phosphate compounds in the orthophosphate monoester group identified by the researchers on a soil sample (0-5 cm) from a plot fertilised under conventional tillage.

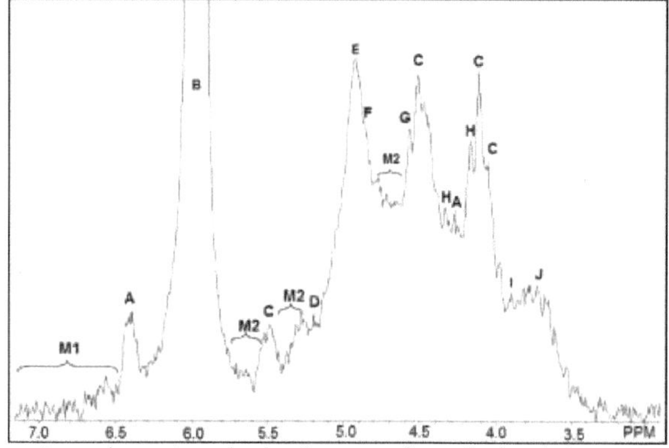

Figure 1-3. NMR spectroscopic spectrum showing the group of organic phosphate compounds detected in the monoester region of a soil sample (0-5 cm) from a plot fertilised under conventional tillage. (A) *neo-IP6*. (B) Orthophosphate. (C) *myo-IP6*. (D) Glucose-6P. (E) Unknown compound. (F) a- Glycerophosphate. (G) b-Glycerophosphate. (H) Nucleotides. (I) Choline-P. (J) *scyllo-IP6*. M1, monoester 1; M2, monoester 2 (Source: Abdi et al., 2014).

All these forms of organic P have been identified using various analytical techniques such as 31P NMR and enzymatic hydrolysis (Condron et al., 2005; Cade-Menum et al., 2010; Abdi et al., 2014). Organic P also exists in the form of compounds that are soluble in the soil solution. However, their contribution to plant nutrition remains uncertain. Soluble organic compounds derived from organic P play an important role in the environment by contributing to the eutrophication of surface waters (Messiga, 2010).

Soil microbial P is highly variable depending on soil typology. In fact, it could represent an average of 2 to 10% of total soil P, depending on the different stages of soil development (Richardson and Simpson, 2011). It could even be as high as 50% in the litter layers of the soil surface (Oberson and Joner, 2005; Achat et al., 2010). In addition, microbial P constitutes a highly dynamic pool of soil P because it is subject to major changes in response to several environmental factors, such as soil temperature, moisture and carbon availability.

P from the biomass is released in the form of orthophosphate ions and in organic forms that are rapidly mineralised in the soil (Macklon et al., 1997). In soils with low adsorption capacity, the release of microbial P could be associated with a measurable increase in soil solution P (Oehl et al., 2001).

1.1.3. Methods for determining the P available to plants

Soil P availability is defined as the quantity of P that can be absorbed by a plant. It is governed by root uptake and soil P reserves (Hinsinger et al., 2007). Soil P availability is also influenced by the physico-chemical and biochemical processes that take place in the soil rhizosphere. The soil solution is the site of P uptake. The species absorbed by the roots are orthophosphate ions ($H_2PO_4^-$ and HPO_4^{2-}), depending on the variation in soil pH. This absorption leads to a drop in the concentration of orthophosphate ions in the soil solution. The soil solution is therefore replenished from the solid phase as a result of the concentration gradient established between the two phases.

Furthermore, soil solution plays a minimal role in phosphate nutrition. In fact, several studies have shown that around 99% of the P taken up by plants during a crop cycle comes from the solid phase (Grant et al., 2005; Morel, 2007). Thus, the transfer of P from the solid phase to the solution occurs by diffusion and numerous associated rhizospheric mechanisms may be involved in this phase change (Hinsinger, 1998; Hinsinger et al., 2005). Messiga (2010) defines three types of interaction between the root-solution-soil system. Firstly, the reactions that determine and control the transfer of P ions between the solid phase of the soil and the soil solution (adsorption/desorption, precipitation/dissolution, diffusion, mineralisation, complexation and redox reactions). Next come the mechanisms linked to root function (presence of mycorhyzogenic fungi and other root exudates). Finally, there are the interactions between soil-solution reactions and the mechanisms implemented by the roots, interactions between root mechanisms and interactions between soil-solution reactions.

Because of the great complexity of the root-solution-soil system, the amount of P actually available to the crop is still difficult to assess. Frossard et al (2004) define crop P availability as the total quantity of P in the soil likely to end up in the soil solution in the form of orthophosphate ions over the growing period of the crop, since plants assimilate P mainly by the root route in the form of orthophosphate ions in solution.

The availability of P to plants is characterised by three factors: intensity, quantity and capacity (Frossard et al., 2004). The *intensity* factor represents the activity of P in solution, completely and immediately available to plants, and is expressed in mg P/l of solution. Thus, P extracted with water or from a diluted salt is used to estimate the *intensity* factor. The *quantity* factor is defined as the total quantity of P likely to pass into solution and become available to plants. In other words, it is the quantity of phosphate ions present on the solid phase and capable of replenishing the solution after phosphate ion uptake by plants (McGecham, 2002; McGecham and Lewis, 2002; Grant et al., 2005; Redding et al., 2006). It is expressed in mg P/kg of soil. Finally, the *capacity* factor represents a soil's ability to keep the intensity factor constant when the quantity varies, either when fertiliser is applied or when it is taken up by plants. Also known as buffering capacity, capacity is expressed in l/kg of soil. The physico-chemical properties of soils (i.e. pH, organic matter content, exchangeable Al) determine the mechanisms that govern these three factors (Tran and Giroux, 1990).

Several methods are used to determine the quantity of P available to plants, including (1) isotopic methods, (2) chemical extraction methods, (3) methods linked to the chemical fractionation of P in soils, (4) spectroscopic methods, and (5) anion exchange membranes.

Isotopic methods

Isotopic methods are based on the use of radioactive P isotopes (^{31}P, ^{32}P and ^{33}P). They are based on the assumption that P isotopes behave identically during chemical, physical and biological processes. Two types of isotopic methods exist: one follows the transfer of radioactive P in solution in the soil to the solid phase in an agitated soil-water suspension (isotopic dilution or isotopic exchange kinetics) and the other, based on labelling the available P in a soil with radioactive P and tracking it to the test plant (Frossard et al., 2004).

Applied mainly in research, isotopic methods give relatively reliable results on the concentration of P in solution, the ability of the soil to buffer P variations in solution and the quantity of P that can arrive in solution over a period of less than three months. However, these methods are not sufficiently accurate for estimating available P levels in soils with high fixing capacity (Tran et al., 1988). In addition, the analytical difficulties associated with using this method in the laboratory make it more complex.

Chemical extraction methods

Chemical extraction methods are the most commonly used methods for determining the P available in soils. These methods make it possible to characterise the P available to plants using acids or bases. According to Tiessen and Moir (2007), the choice of a chemical extraction method is based on four main factors, such as (1) the simplicity of the method, (2) sufficient extraction of P available to plants for measurement purposes, (3) the significant proportion of extracted P that is exported by plants, and (4) the risk of extracting P that is not available to plants during the growth period. Thus, some methods will tend to extract more easily P attached to soil particles, while others will represent more P in solution. Pedogenesis and soil properties (soil type, water pH, clay content, organic matter content) have a major effect on the results of the various tests available. Table 1.1 summarises the main methods used for the chemical extraction of available P.

Table 1.1 - Chemical extraction methods for available soil P

Method	Chemical composition	Soil : solution ratio	Time
Mehlich-3 (Mehlich, 1984)	$0.015N\ NH_4F + 0.25N\ NH_4NO_3 + 0.2N\ CH_3COOH + 0.013N\ HNO_3 + 0.001M$ EDTA at pH 2.3	1:10	5 min
Mehlich-2 (Mehlich, 1978)	$0.2N\ NH_4Cl + 0.2N$ HOAc + $0.015N\ NH_4F + 0.012N$ HCl	1:10	5 min
Mehlich-1 (Nelson et al., 1953)	$0.05N$ HCL + $0.025N\ H_2SO_4$	1 :4	5 min
Bray 2 (Bray and Kurtz, 1945)	$0.03N\ NH_4F + 0.1N$ HCL	1:10	1 min
Bray 1 (Bray and Kurtz, 1945)	$0.03N\ NH_4F + 0.025N$ HCL	1 :10	1 min
Olsen (Olsen et al., 1954)	$0.5M\ NaHCO_3$ at pH 8.5	1 :20	30 minutes

Adapted from Giroux and Tran (1985), Tran et al (1990) and Beagle (2005)

In North America and Canada in particular, the most widely used methods are the Mehlich-3 method (Mehlich, 1984) for acid soils and the Olsen method (Olsen et al., 1954) for alkaline soils. In Quebec, the Mehlich-3 method has been used since 1986. This chemical extraction method has the advantage of simultaneously extracting several mineral elements, such as P, Mg, Ca, Na and K, as well as the following trace elements: Al, Cu, Zn, Mn and Fe. It is a robust method that has been tested and found effective on the acidic and neutral soils of Quebec.

Agri-environmental indicator of P

Studies have made it possible to significantly improve the use of methods for chemically extracting P from the soil for the purposes of recommending phosphate fertilisers that take agronomic aspects and environmental risks into account. Based on the concept of the degree of P saturation developed by Breeuwsma and Silva (1992) and the Mehlich-3 chemical P test, Giroux and Tran (1996), Khiari et al. (2000), and Sims et al. (2002) have developed an agri-environmental P indicator for acid soils in Quebec and the United States, respectively.

The agroenvironmental P indicator (or soil P saturation index (PSI)) is defined as the proportion of fixation sites already occupied by P. Aluminium and iron are the main fixation sites, although calcium-bound forms of P are dominant in calcareous and neutral soils in Quebec (Giroux and Tran, 1985; Tran and Giroux, 1985). The PSI is therefore based on the ratio (Mehlich-3 phosphorus)/(Mehlich-3 aluminium), hence the $[P/Al]_{M3}$ ratio. Based on this $[P/Al]_{M3}$ ratio, Giroux and Tran (1996) evaluated P saturation (SP) in Quebec soils in 4 agroenvironmental risk classes: (1) low saturation (0% <SP < 5%), (2) medium saturation (5%<SP<10%), (3) high saturation (10%<SP<20%) and (4) very high saturation (SP>20%). Pellerin et al (2006) adapted the agri-environmental P indicator to maize growing in Quebec. Ige et al. (2005) used the Olsen chemical extraction method to develop an agri-environmental P indicator specific to the neutral and calcareous soils of Canada's northern plains. In Quebec, the critical agri-environmental threshold $(P/Al)_{M3}$ for fine-textured soils corresponds to a limit value of 8%, while this value corresponds to a limit threshold of 15% for coarse-textured soils (CRAAQ, 2010). These critical P saturation indices have been incorporated into the Quebec government's environmental regulations on the sustainable management of agricultural land (MDDELC, 2017).

Chemical soil extraction methods are the most commonly used methods for determining the P available in soils, but these methods are time-consuming, laborious, environmentally unfriendly and result in the destruction of soil samples.

Chemical fractionation methods

Labile soil P is defined as P adsorbed on the surface of soil particles, which is in rapid equilibrium with phosphate ions in the soil solution. In other words, it is the proportion of P that is easily mobilised. Depending on its availability to plants, labile soil P could be distributed in three pools or fractions (including organic or inorganic P), namely (1) the easily labile pool, (2) the moderately labile pool and (3) the pool that is difficult to labile (Tiessen et al., 1982; Zheng and Zhang, 2011). Soil labile P fractionation has applications in the development of chemical P fractionation methods.

Sequential P extraction methods were initially developed, then adapted and modified to improve the extraction of inorganic and organic P fractions in cultivated soils (Chang and Jackson, 1957; Fife, 1963; Petersen and Corey, 1966; Williams et al., 1967; Smillie and Syers, 1972; Bowmen and Cole, 1978). However, these methods had limitations in that they determined the inorganic and organic fractions of P non-simultaneously.

Thus, an improved P fractionation protocol was developed by Hedley et al (1982) to simultaneously determine the inorganic and organic fractions of P. The protocol is summarised in 3 phases. Phase 1 determines the available and easily labile P using anion exchange resin, followed by 0.5M NaHCO3 (P_i/P_o). This step assesses the P readily adsorbed by carbonates, sesquioxides and crystalline minerals (Zheng and Zhang, 2011). Phase 2 determines the moderately labile P or "protected P" contained in the particles (amorphous Fe and Al oxides) using 0.1M NaOH (P_i/P_o). Apatite minerals (Ca-bound P) are extracted using 1M HCl (P-HCl) during this phase. Finally, phase 3 determines the residual P using a H2S04-H2O2 (or $K_2S_2O_8$) digestion.

The Hedley method has been widely used to assess the distributions of P fractions in cultivated agricultural soils (O'Halloran, 1993; Piegholdt et al., 2013; Shi et al., 2013), however modifications have been made to easily extract labile P (Tiessen and Moir, 1993; Kuo et al., 1996; Piegholdt et al., 2013; Sun et al., 2015). P fractionation methods are more descriptive than chemical analysis methods, but they are more time-consuming to perform routine soil analyses. In addition, the recovery rate of the different P fractions analysed using these methods varies widely compared to other methods (Ziadi et al., 2013).

<u>Spectroscopic methods</u>

Spectroscopic methods include the near infrared spectroscopic technique, nuclear magnetic resonance (NMR), and techniques based on the XANES method (synchrotron radiation). Initially used in the laboratory, the near infrared spectroscopic technique is now being applied in the field. The NIR spectroscopic method is fast, accurate, inexpensive and environmentally friendly (Ndwamungu et al., 2009a; 2009b; Abdi et al., 2016). The method requires little preparation, no logistics and no chemicals to analyse soil samples. The near-infrared spectroscopic technique is based on calibration equations between the optical signal and the physico-chemical parameter being measured, which is why it is so important to carry out sample collection campaigns in order to obtain a database.

Several studies have shown that the near-infrared spectroscopic technique is effective for the determination of total P in fine clayey and sandy soils (Malley et al., 2004; Bokregci and Lee; 2005), and organic P in loamy and loamy clay textured chernozems (Abdi et al., 2016). The success of the prediction quality of the organic fraction of P measured by this method relies on the fact that organic compounds are more easily excited by infrared radiation than inorganic fractions of P. However, this method is less effective for the prediction of available P extracted at Mehlich-3. Chang et al (2001) and Nduwamungu et al (2009a) showed a poor correlation between available P measured with Mehlich-3 and the spectroscopic technique (VNIRS) for soil samples collected from 448 sites in the USA and 150 samples from 5 soil series in the Montreal area. However, this method provided good results for the determination of available P using the Olsen method (Van Groenigen et al., 2003; Abdi et al., 2016). Near infrared spectroscopy therefore remains a very promising method for the future (Abdi et al., 2016).

NMR spectroscopy and the XANES method have also been used for more precise identification and quantification of P forms in soils (Cade-Menum et al., 2010; Abdi et 2014). These techniques have been described in detail by Beauchemin et al. (2003) and Shober et al. (2006). However, the application of these techniques is very limited due to the financial costs and the technical and practical difficulties associated with their use.

<u>Anion exchange membranes</u>

Anion exchange membranes (AEMs) are also widely used to assess the P available to plants. They simulate root action by continuously removing P from solution (Yang et al., 1991), which integrates the effect of biotic and abiotic soil factors (Cooperband and Logan, 1994). By acting as a sink for orthophosphate ions, MEAs allow better determination of the soil *quantity* factor than chemical extraction methods. As a result, they improve measurements of soil buffering capacity and P sorption and desorption curves (Sato and Comerford, 2006).

Several studies have shown that MEAs or root simulators could be better indicators of the availability to plants of soil nutrients such as nitrogen and soil P (Quian et al., 1992; Ziadi et al., 2000; Nyraneza et al., 2011; Mason et al., 2013). To this end, a study carried out in Quebec on a humic Inceptisol showed that P extracted using a MEA is a better indicator of soil labile P than P extracted using the Mehlich-3 chemical test (Zheng et al., 2003). Furthermore, Zehetner et al, (2018) observed that MEAs correlated better with plant P compared to the Mehlich 3 method in a comparative study of 14 P extraction methods performed on 50 Austrian and German agricultural soils grown with spring wheat. Furthermore, the use of MEAs incorporates the temporal variation linked to the replenishment of P over a certain period, compared with chemical extraction methods, which are static measurements of phosphate ions at one point in time (Nyraneza et al., 2011).

MEAs could therefore be a better alternative and an excellent potential tool for determining available P than chemical extraction methods, which are costly, time-consuming and environmentally unfriendly for analysing this major element. However, although it has been used successfully in conventional agriculture, the use of MEAs to determine available P is still very limited, particularly in soils under arable crops.

1.2. Agronomic practices and environmental factors related to P management in cultivated soils

1.2.1. Tillage and P in agricultural soils

The sustainable management of P in field crop soils is based on the application of appropriate phosphate fertilisation recommendations combined with suitable cropping management practices (Messiga, 2010). Conventional tillage (CT) is a cultivation practice based on the formation of a plough layer using a share (or disc) plough. Soil tillage controls weeds, modifies the physical properties of the soil (porosity and drainage) and improves crop productivity in conventional farming systems. However, CT destroys soil structure and increases the risk of soil erosion, which increases the risk of nutrient loss (Li, 2017).

Soil conservation farming practices (i.e. minimum tillage or direct seeding (DS)) have been developed to promote sustainable farming systems. SD is a better soil conservation practice than CT because it increases soil biological activity (Duiker and Beegle, 2006; Abdi et al., 2014) and improves soil physical properties by reducing compaction, drainage risk, water and wind erosion risk in agricultural soils (Dick, 1992; Olson and Ebelhar, 2009; Friedrich et al., 2012).

SD is practised on 9% of arable land worldwide, or 125 million hectares, particularly in North America (Pittelkow et al. 2015). In Canada, the percentage of arable land under SD increased from 18.5% in 2001 to 25% in 2011. In 2016, Canadian farmland under SD covered an area of 19.5 million hectares (Statistics Canada, 2016). In Quebec, the area of land under SD doubled to more than 69% of the total area between 2009 and 2013 as a result of increasing government subsidies to farmers to promote SD (Statistics Canada, 2016). These increases in area under SD were observed because of good agronomic and economic performance (Holm et al., 2006; Lafond et al., 2011).

SD influences the chemical properties of agricultural soils, particularly at the level of P. Indeed, P contents are very high in the first 5 cm of the soil, then decrease drastically at depth in systems under SD, compared to those under CT (Duiker and Beegles, 2006; Cade- Menum, et al., 2010; Messiga et al., 2012; Rodrigués et al. 2016). Abdi et al. (2014) found high levels of total P and available P (P_{M3}) in the (0-5 cm) and (5-10 cm) horizons in soils under SD compared to soils under CT in their study on the impact of tillage and phosphate fertilisation on the distribution of P forms in the soil. The researchers also observed a distribution of total carbon similar to available P in soils under SD compared with soils under CT.

Rodrigués et al. (2016) also observed an increase in organic and inorganic P forms in the surface horizons (0-5 cm) of 4 tropical soils (Oxisols type) under SD compared to soils under CT, following repeated applications of phosphate amendments. Cambouris et al. (2017) observed similar results regarding the distribution of available P (P_{M3}) in the surface horizon (0-5 cm) of soils under SD compared to soils under CT. Indeed, P_{M3} contents and PSI were higher in soils under SD compared to soils under TC in long-term plots under maize-soybean rotation. Furthermore, Puustinen (2005) also showed that dissolved reactive P transport was 348% higher in soils under SD compared to soils under CT.

Thus, SD increases the risk of accumulation of available P, particularly in the surface horizons (0-5 cm). Abdi et al. (2014) refer to P stratification to define P accumulation in surface horizons under minimum tillage. SD could be a significant agri-environmental issue due to a high risk of P transport and leaching into water.

1.2.2. Stratification of P linked to the agronomic practice of direct seeding

P stratification in soils under SD is characterised by high P concentrations in the first five centimetres of soil, decreasing with soil depth (Selles et al., 1999; Duiker and Beegle (2006); Cade-Menun et al., 2010; Messiga et al., 2012; Abdi et al., 2014). P stratification is thought to be mainly caused by three reasons (1) the annual broadcast (or banded) application of phosphate amendments close to row seedlings, (2) the lack of mixing of phosphate amendments (fertilizers, manures) with soil and crop residues left on the soil surface after harvest, and (3) the subsequent leaching of P from crop residues (Duiker and Beegles, 2006; Piegholdt et al., 2013).

To overcome this, two strategies have been proposed, namely (1) strategic minimum tillage or reduced tillage (Dang et al., 2015) and (2) intensive soil sampling (Cade-Menum et al., 2010). Strategic minimum tillage could improve the profitability and sustainability of farming systems in the short term, particularly for P. Indeed, Dang et al. (2015) argue that strategic minimum tillage combined with SD could reduce the stratification of available P in the top 10 cm of the surface layer, and would help redistribute available P deeper into the 0-30 cm layer for at least three years. However, in the long term, the impact of this cropping practice could be negative at the agri-environmental level, particularly for soil carbon and nitrogen cycles (Dang et al., 2015).

Intensive soil sampling, based on the collection of soil samples from different horizons, could provide a better understanding and identification of the vertical variability of P in the root zone for the purposes of precise P fertilisation recommendations (Cade-Menum et al., 2010). This georeferenced sampling strategy would make it possible to identify excessive and critical P zones at field level.

1.2.3. Environmental factors and P in agricultural soils

Environmental conditions (freeze/thaw cycles, desiccation/humidity and other variations in abiotic factors) could also modify P levels in agricultural soils. Cold temperate soils are exposed to freeze-thaw cycles, which could stimulate mineralisation and affect soil P availability (Freppaz et al., 2007). The destruction of plant cells releases soluble P from plant residues (Bechmann et al., 2005).

Messiga et al (2010) investigated the effects of freeze-thaw cycles on layer (0-15 cm) P availability of soils grown under SD compared to soils under CT under field and laboratory conditions during an increasing number of 1-6 freeze-thaw cycles. The researchers observed that water-soluble P (P_W) and available P (P_{M3}) contents were very high in soil samples collected in spring compared with those collected during autumn. In addition, between the first and third freeze-thaw cycles, mean soluble P concentrations were twice as high under SD compared with TC. After 6 alternative cycles of freezing and thawing, the accumulation of available P was particularly higher in soils under SD compared with soils under TC because P_{M3} contents had doubled.

Thus, soluble P (P_W) and P_{M3} contents could be elevated in soils under SD compared to soils under CT, following an increase in the number of freeze-thaw cycles under global warming conditions. This should increase soil available P concentrations in early spring, however it remains to be determined whether soluble P would benefit early season crop growth or be lost to runoff and drainage during spring thaw.

Furthermore, soil water content varies rapidly after rainfall in several agroecosystems (Ziadi et al., 2013). The resulting desiccation/rewetting cycles could therefore affect available soil P levels. The increase in P available in the soil during the rewetting phase results from the release of organic P from the lysis of microbial biomass cells (Turner and Haygarth, 2001). Soil rewetting is accompanied by rapid microbial growth and respiration, probably due to new microbial cells using soluble substrates, such as soluble organic P compounds released during the drying phase. Barrow and Shaw (1980) demonstrated that the effects of desiccation (60°C) on the concentration of available P in the soil were short-lived, as the concentration of soluble P returned to the pre-treatment level after 1 day of soil rewetting.

Moreover, sustainable P management in cultivated soils depends on balanced P fertilisation (Tunney, 1990; Fu et al., 2013). One of the major ways of achieving this objective is to have a better understanding of the spatial variability of P in field crop soils (Haneklauss and Schnug, 2006).

1.3. Spatial variability of soil P in cultivated soils

1.3.1. Soil variability: definition and concepts

Several authors (Wagenet, 1985; Ovalles and Collins, 1986; Nolin and Caillier, 1992a; b; Si et al., 2007; Franzen,

2018; Kitchen and Clay, 2018) have explored the concept of soil variability. Kitchen and Clay (2018) summarised soil variability as the measure of dissimilarity or difference, of soil yields, properties or nutrients in space or time, across a field. Previously, Nolin and Caillier (1992a) defined soil variability in three components, namely vertical variability, lateral (or spatial) variability, and temporal variability. According to the authors, vertical variability translates into the differentiation of the soil profile into distinct horizons or layers, lateral (or spatial) variability is characterised by variations in the cultivated surface layer, while temporal variability is associated with daily, weekly, monthly or seasonal variations in the soil variables or descriptors measured. Temporal variability is often masked by vertical and lateral variability (Cameron et al., 1971; Nolin and Caillier, 1992a).

The spatial and temporal variability of soils can be of intrinsic or extrinsic origin (Wagenet, 1985; Franzen, 2018). Intrinsic variability is linked to natural soil-forming factors (origin of parent material, types of soil series), while extrinsic variability is linked to anthropogenic factors associated with land management (cultivation practices, intensive fertilisation practices). The interaction of intrinsic and extrinsic factors also contributes to soil variability (Nolin and Caillier, 1992a). In short, spatial variability represents the amplitude of variation in a soil variable studied at plot level (Arvalis, 2015).

Spatial variability (SV) can be studied using the classical statistical approach combined with the geostatistical approach (Nolin and Caillier, 1992b; Arrouays et al., 1997; Rivero et al., 2012; Mondal et al., 2020; Zhang et al., 2021). The statistical approach, based on the study of classic descriptive statistics (minimum, maximum, mean, standard deviation and above all the coefficient of variation (CV)) assesses the degree of heterogeneity and dispersion of the soil descriptor studied. The geostatistical approach analyses the spatial structure of the soil descriptor or variable studied, based on the spatial dependence of the samples analysed.

The Variogram is the basic tool in the geostatistical approach. It represents the dissimilarity between points as a function of the distance between them. By studying the Variogram, it is possible to assess the importance of the SV of soil properties. The presence of spatial structure in soil properties can be determined from the main components of the Semivariogram, namely the random variance or nugget effect (c_0), the structured variance (C), the plateau (c_0+c) and the range (A) summarised in Figure 1.4.

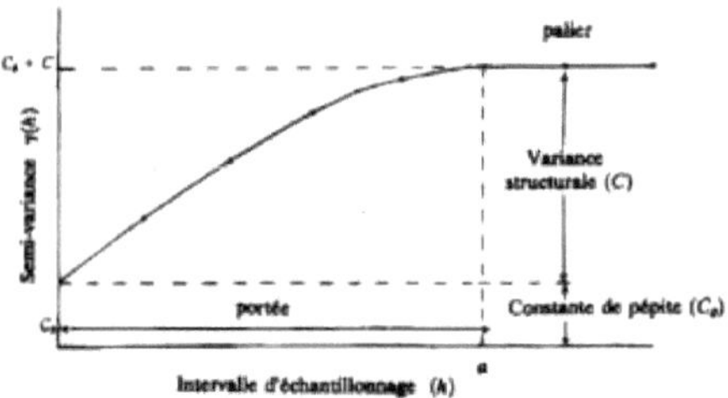

Figure 1-4. Representation of a semi-variogram and its components

(Source: adapted from Nolin and Caillier (1992b))

Thus, the partial threshold ratio (R=C/(c_0+ C)), expressed as a percentage of the total semivariance, defines the proportion of spatial dependence or structure of the soil property analysed. Semivariograms with a partial threshold ratio < 25%, 25-40%, 40-60%, 60-75% or > 75% correspond to weak, weakly moderate, moderate, strongly moderate, or strong spatial dependencies, respectively (Whelan and McBratney, 2000). Previously, Cambardella et al (1994) inversely defined this nugget/threshold ratio ($R=c_0$/(c_0+c)) in three classes in order to define the proportion of spatial dependence of the parameter analysed. A ratio of R<25%, 25-75% and >75% indicates strong, medium or weak spatial dependence, respectively. This scale is also used in many studies.

The degree of spatial dependence (R) of a parameter is related to intrinsic or extrinsic soil factors. Indeed, Vasu et al (2017) state that parameters with a high spatial structure are influenced by intrinsic soil factors (soil types, soil formation factors) while parameters with a low spatial structure are influenced by extrinsic factors (cultivation practices, use of soil improvers and fertilisers).

1.3.2. Spatial variability of P on a territorial, regional and catchment scale

Several authors have studied the SV of P in cultivated soils ranging from a territorial to a regional scale (Jia et al., 2011; Roger et al., 2014; Wei-Xia et al., 2015; Vasu et al., 2017), and catchment scale (Nolan et al., 2007; Yuan et al., 2009; Wilson et al., 2016; Zhang et al., 2021). Studies on soil P SV at territorial and regional scales have generally been carried out for planning purposes and environmental P mapping in long-term regions (Gburek et al., 2000; Klatt et al., 2003; Roger et al., 2014).

Some studies of P SV aimed to map the P available in soils for environmental pollution prevention purposes. Specifically, Roger et al (2014) mapped soil P in the canton of Fribourg (Switzerland) in order to identify sources of P available in the region for environmental pollution prevention purposes. The results showed that the SV of P is significantly related to the type of land use in the Fribourg region in terms of total P content, available P and the degree of P saturation (DSP). The researchers found that soils cultivated under maize-wheat rotation had the lowest total P content (935 ppm), but conversely had the highest available P content (81.3 ppm) and degree of P saturation (37.9%), compared with soils under grassland (67.6 ppm) and pasture (22 ppm). In other words, the highest concentrations of available P were in cultivated soils, followed by soils under grassland, and soils under mountain pastures.

Wei-Xia et al. (2015) investigated the SV of P in intensively cultivated soils in the Yangtze River Delta region (China) to assess the degree of soil P pollution, and to issue future agricultural strategies to reduce P concentrations in the region. The results showed that the SV of available P extracted using the Olsen method (P_{Olsen}) was strongly linked to intrinsic factors, in particular the different soil series encountered in intensively farmed soils. Indeed, strong spatial dependencies (R>84.5%) were observed by the researchers.

Vasu et al (2017) studied the SV of P in the arid tropical region of the Deccan Plateau (India) to identify critical areas deficient in major nutrients, including available P (P_{Olsen}). The Deccan Plateau covers an area of 0.42 million km^2 . The results showed high variability (CV=59%) and moderate spatial dependence of P (R=27%). In addition, maps of the spatial distribution of nutrients produced by ordinary kriging interpolation were used to identify P-deficient areas in the region, ranging from 4 to 12 kg ha^{-1} . The sampling grid used was 325 m x 325 m.

The objective of watershed-scale P SV studies was to assess the spatial distribution of available P in cultivated soils for the purpose of predicting soil erosion risk, leading to dissolved P losses. For example, Wilson et al. (2016) investigated the SV of available P (P_{Olsen}) in 16 small watersheds in southwestern Manitoba draining eight fields in relation to landform, tillage, and P inputs. The objective of the study was to investigate the spatial distribution of Olsen P at the watershed scale in order to assess the impact of different cropping practices, relief and topography on the spatial variability of Olsen P$_{sem}$ The results showed that the variability of measured available P (Olsen P) was a function of relief and topography in relation to soils under different cropping practices. To this end, the researchers observed significant correlations between the Olsen P levels analysed and the topographical moisture index, ranging from 0.19 to 0.96 (r^2 =0.25, n=56, p<0.0001), for the purposes of predicting a high risk of P losses in the two horizons studied (0-15 cm) and (15-60 cm). In addition, extractable P levels were higher in soils with lower slopes than in soils with higher slopes. The researchers explained this by the fact that P is strongly fixed by Ca in soils with high slopes, making P less available. However, the researchers suggested that further work should be carried out to develop quantitative relationships between the extractable P levels analysed and the cropping practices specific to the watersheds and hydrological processes dominant in the south-western region of Manitoba.

1.4. Spatial variability of P at farm plot level

The study of soil P SV at the scale of an agricultural plot, is a very important approach, as it represents an effective tool for sustainable P management in precision agriculture (Fu et al., 2013a; Kitchen and Clay, 2018). Indeed, agricultural producers are interested in managing soil variability within and beyond a field, as this allows them to improve land productivity (yield assessment), increase business profitability while improving environmental protection by reducing P losses (Mallarino et al., 2007; Ferguson and Hegert, 2009).

However, to our knowledge, few studies have been carried out specifically on the SV of P at the scale of an agricultural plot in field crop soils. This section therefore summarises the few studies that have been carried out on the SV of P under the effect of field crops and cropping practices.

1.4.1. Spatial variability of P linked to cropping systems and land use

The SV of P (total P, available P) in soils under arable crops is still poorly understood. Indeed, Fu et al (2013a) carried out a study on the SV of four physico-chemical soil properties (pH, Mg, K and P measured by the P-Morgan) of two fields under silage and pasture crops in Ireland, following repeated applications of poultry litter. The aim of this study was to describe and understand the SV of P at the two sites using statistical and geostatistical tools. Field 1 had been under silage for 15 years and covered an area of 5.2 ha, while Field 2 had been under permanent pasture for 16 years and covered an area of 9 ha. The results showed significant positive correlations between soil P and the other soil properties studied (pH, K and Mg) at both sites. The researchers also observed that the coefficient of variation (CV) of P at site 2 (87.4%) was higher than at site 1 (28.8%), reflecting the high intensity of variation of P at site 2 compared with site 1. In addition, site 2 had a lower spatial dependency ratio for P (and other nutrients) than site 1, reflecting the low spatial structure of P and the other soil properties studied at site 2 compared with site 1. Thus, the authors deduced that permanent sheep grazing on site 2 could decrease the autocorrelation of soil nutrients, which could demonstrate the impact of cropping practices (extrinsic soil factors) on modifying the spatial structure of site 2.

Furthermore, the SV of P (total P, available P and DSP) could also be influenced by soil use. Bennett et al. (2005) observed a high variance (expressed in log (mg kg)$^{-2}$) of available P (P-Bray) in soils under grain maize (0.61) compared with soils under grassland (0.12) in the state of Wisconsin. Similarly, Roger et al (2014) observed higher available P variability in soils under grassland (CV=106.7%), followed by soils under mountain pasture (CV=86.6%), and cultivated soils under maize-wheat rotation (CV=69.7%) in the canton of Fribourg (Switzerland). In addition, the researchers observed that concentrations of available P (ppm) and DSP (%) were significantly higher in cultivated soils, followed by soils under grassland and soils under mountain pastures. In fact, the average levels of available P analysed (P extracted using the ammonium acetate-EDTA (P_{AAE}) method: the Swiss reference method) were 81.3 ppm for cultivated soils, compared with 67.6 ppm for grassland soils and 71.3 ppm for mountain pasture soils. The DSPs analysed were 38%, 31% and 17% respectively. Conversely, Roger et al (2014) observed that the highest total P concentrations were found in soils under mountain pasture, followed by soils under grassland and cultivated soils, and came from the highest levels of P_{org} analysed, which would confirm the contribution of organic amendments.

In terms of geostatistical analyses, the researchers observed a spatial dependence of P indicators greater than 70%, resulting from the influence of the high soil diversity encountered in the region. To better address the problem of spatial dependence, Roger et al. (2014) suggested an additional regular and georeferenced sampling strategy for a better spatial analysis of P indicators in relation to intrinsic and extrinsic factors but also for a reduction in the uncertainty of P prediction models.

1.4.2. Spatial variability of P in relation to cropping practices

Tillage combined with cropping practices (extrinsic factors) such as P-rich mineral and organic amendments (and their history), composts, incorporation of crop residues into the soil, crop rotation and land use significantly affect the spatial distribution of available P in cultivated soils (Bennett et al., 2005; Page et al., 2005; Nolan et al., 2007; Wei-Xia et al., 2015).

For example, Wilson et al. (2016) characterized the SV of P_{Olsen} in relation to five different cropping practices, combined with tillage (no-till with commercial fertilizer application, organic site without amendment, site with organic slurry amendment, site with historical and repeated slurry application, and soils under conventional tillage with commercial fertilizer use) in their study of Manitoba soils according to depths (0-15 cm) and (15-60 cm). The results showed marked differences in the impact of cropping practices on P management. In fact, the highest levels of P were observed on sites that had historically received repeated applications of fertiliser or slurry with ploughing (20-22 ppm P), in contrast to no-till soils, and those that had received little or no organic amendments (4.5-6 ppm P). The researchers also observed that there was very marked variability in the P content of the surface horizons (0-15 cm) of soils with low P content (organic soils receiving no soil improvers, and no-till soils), compared with the surface horizons (0-15 cm) of soils enriched with P (conventionally tilled soils, soils receiving commercial fertilisers, soils historically receiving liquid manure). However, this P variability linked to cropping practices was less marked in the sub-surface horizons (15-60 cm) of the two soil groups analysed.

In addition, the SV of P (total P, available P) under the influence of tillage (TC compared with SD) remains little studied. Cambouris et al. (2017) studied the decimetric SV of available P (Mehlich-3, $_{PM3}$), on grain maize rows of experimental plots influenced by two different cropping practices (conventional tillage and no-till) using conventional chemical and 2D geostatistical analyses (semivariograms), in the (0-5 cm) and (5-20 cm) horizons. The plots studied were fertilised with 35 kg P ha^{-1} every two years in a long-term experimental site under maize-soya rotation. The researchers observed that the horizontal distribution of $_{PM3}$ was less sensitive to extrinsic factors, notably tillage, P fertilisation and soil depth. In addition, the distribution of $_{PM3}$ in 2D (horizontal and vertical) according to soil depths (0-5 cm) and (5-20 cm) on either side of the row seedings did not reveal a VS of the soils influenced by the 2 cultivation practices studied. According to the researchers, the high variability, observed by the high CVs (77% and 63%), associated with $_{PM3}$ levels in the plots under CT and SD could have overshadowed the strong decimetric distribution of P marked in the form of variability. As the study was carried out and limited to the plot scale, the 2D geospatial model linked to the fertilisation method was not detected by the sampling grid used. Thus, Cambouris et al. (2017) suggested developing a geostatistical sampling procedure that is better equipped and more appropriate at the scale of an agricultural field in order to obtain a better estimate of P variability in a soil under SD.

To sum up, despite the studies carried out and mentioned above, knowledge of the SV of available P in arable soils, under the effect of cropping systems and tillage (CT vs. SD) is still very limited.

1.5. Management of spatial variability of P on the scale of an agricultural plot

1.5.1. Management zone: a tool for managing the spatial variability of P

Managing variability for economic return, sustainable development is the basic principle of precision agriculture (PA) (Kitchen and Clay, 2018). Understanding P SV represents an effective strategy for sustainable P management in PA insofar as it enables rational use of phosphate fertilisers. This would reduce environmental P losses through erosion and runoff.

Controlling soil P SV under field crops in PA is based mainly on two main agronomic approaches: (1) variable rate application (VRA) and (2) the use of management zones (MZs). VRA manages the variability of a parameter on the basis of a prescribed interpolation map produced using georeferenced data combined with a geostatistical analysis (*i.e. a* kriged map of the variability of NO3-N concentrations in the soil).

The ZAs approach is an increasingly popular agronomic strategy in PA (Fergusson et al., 2002). It consists of subdividing agricultural plots into small homogeneous soil units based on their characteristics and behaviour in terms of specific needs and responses to inputs and environmental risks in relation to targeted crop yields (Mulla, 1989; Fridgen et al., 2000, Khosla et al, 2001; Quenum et al., 2012, Cambouris et al, 2014; Servadio et al., 2017). These homogeneous units, called ZAs, are used to capture the SV, guide soil sampling by reducing the number of samples taken and therefore the sampling density, establish recommendations and manage inputs (*e.g.* pesticides, fertilisers, irrigation) uniformly between the different units (Cambouris et al. 2006).

The ZA of a field under cultivation is characterised by a sub-region with a combination of homogeneous factors limiting the targeted crop yield (Vrindts et al., 2005; Li et al., 2007). ZAs have many uses. They represent an excellent alternative to grid soil sampling and are tools for assessing agronomic yields in relation to the physicochemical properties of soils for a better study of crop modelling (Li et al., 2007; Armstrong et al., 2009; Rab et al., 2009). ATV is more effective with plant-based properties, while the ZAs approach is more effective with soil properties than with plant-based parameters (Zebarth et al., 2009).

Several studies have been conducted on the use of ZAs for sustainable N management at a field scale in PA (Khosla and Alley, 1999; Khosla et al., 2002; Cambouris et al, 2006; Peralta et al., 2015). However, to date, few studies have been carried out on the application of this agronomic approach for sustainable P management at the scale of an agricultural plot in Canada. Furthermore, the spatial information used to delimit ZAs must be quantitative, densely measured, temporally stable, and closely related to the crop yield or input to be managed (Doerge, 2011). Therefore, it is necessary to efficiently delineate plot-specific ZAs (Yao et al., 2014).

1.5.2. ZAs delimitation methods

1.5.2.1 ZAs delimitation criteria

Before ZAs are delimited, it is important to characterise the SV of the property under study at field scale using an intensive sampling grid (Arvalis, 2015). The sampling strategy used in this initial approach is systematic grid sampling. Grid sampling aims to identify the SV of the parameter under study using composite, georeferenced, randomised samples.

The choice of sampling grid remains appropriate. Some authors (Cambardella et al., 1994; Cambardella and Karlen (1999); Robinson and Mettermicht, 2006; Bogunovic et al., 2014) recommend smaller sampling grids (15 m x 15 m), particularly in intensively cultivated agricultural soils on small areas (Mallarino et al., 2014), while others (Saglam et al., 2011; Behera and Shukla, 2015; Vasu et al., 2017) propose larger sampling grids. Chang et al. (1999) reported that the impact of grid distance on the reproducibility of SV measurements decreased significantly with grid sizes greater than 30 x 30 m. McBratney and Pringle (1999) suggest instead that a grid spacing of 20-30 m is generally necessary when applying site-specific management at high resolution). Grid sampling relies on relative sampling density to reveal fertility patterns within a field.

However, grid sampling has its limitations. It is costly, intensive and labour-intensive under different cropping practices. In addition, this traditional method is not viable for obtaining ZAs, as it requires a large number of soil samples to obtain a good spatial representation of soil properties and nutrient levels. Mallarino (1996) mentioned that grid sampling strategies misrepresent available soil P and K under direct seeding. There are other geophysical methods such as electromagnetic induction and electrical resistivity (i.e. apparent electrical conductivity of the soil) that accurately and efficiently measure the SV of soil properties for the purpose of delimiting ZAs in PA (Peralta and Costa, 2013; Adamchuk et al., 2015).

1.5.2.2 Apparent electrical conductivity of the soil: a tool for delimiting ZAs

Apparent electrical conductivity (AEC) is one of the best modern technologies for delineating intraparcel ZAs (Cambouris et al., 2006). This method is characterised by its high measurement density, accuracy and speed. ECa can also be measured simply and inexpensively. Moreover, ECa is related to various physical and chemical properties in a wide range of soils (Sudduth et al., 2005), because it depends on 1) the chemical composition of the soil solution and exchangeable soil ions, 2) the clay content and 3) the interaction between non-exchangeable and exchangeable ions (Nadler and Frenkel, 1980; Shainberg et al., 1980; Rhoades et al., 1989, Heiniger et al., 2003).

The importance of soil ECa is based on the principle that sands, silts and clays have low, medium and high conductivities respectively (Lund et al., 1999). ECa maps are therefore important tools for allocating homogeneous zones because of their strong correlation with several physico-chemical soil properties such as texture (Williams and Hoey, 1987), water content (Kitchen et al., 2005), salinity (Heiling et al., 2011), cation exchange capacity (Saifuzzaman et al., 2021), and organic matter content (Becker et al., 2022).

Soil Proximal Sensors (CPS) are tools for measuring soil ECa in PA to complement and replace conventional methods such as intensive soil sampling with the aim of mapping field variability and delineating ZAs. CPSs that measure ECa continuously are increasingly used to acquire detailed, dense and rapid soil data (Corwin and Lesch, 2003, Mertens et al., 2008; Becker et al., 2022). CPSs are coupled with a global positioning system (GPS) to acquire georeferenced data and increase the efficiency of ZAs delimitation (Corwin and Plant, 2005; Perron et al., 2018).

The Veris® proximal sensor, model 3100 (Veris Technologies, Inc. Salina, KS) is one of the most widely used systems for measuring soil ECa based on electrical resistivity. It uses four electrodes that are in physical contact with the soil and inject an electric current to measure the resulting voltage (Lund et al., 1999). These electrodes are distributed in the form of discs assembled with a two-wheeled metal structure, enabling it to be towed by a sports utility vehicle (Figure 1.5). The further apart the discs are, the deeper the electric current penetrates. The system is designed to record ECa simultaneously at two soil depths, 0-0.3 m and 0-1.0 m, called ECa_{30} and ECa_{100} respectively. ECa is measured in millisiemens per metre (mS m^{-1}). The Veris® model 3100 is combined with a GPS receiver to ensure the precise location of each recorded measurement (Konstantinos et al., 2007). It can therefore be used to measure soil variability by directly detecting electrical conductivity (Lund et al., 1999).

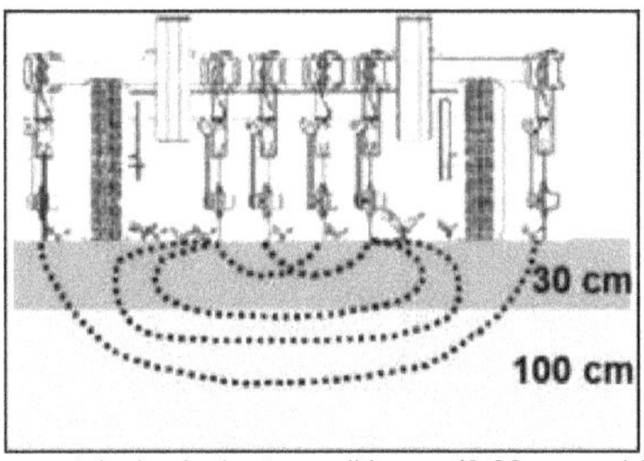

Figure 1-5. Soil ECa measurement device in the two soil layers (0-30 cm and 0-100 cm) (Lund et al., 1999).

Using the Veris®, model 3100 has several advantages such as initial programming and no need for calibration for each application. ECa measurements are not affected by natural magnetic fields, metal objects, or nearby engines (Adamchuk et al., 2015). Nevertheless, the Veris 3100 is heavy and requires a tractor to pull it in the field. Its use is also limited by soil moisture conditions (Quenum et al., 2012) and crops (Sudduth et al., 2003; Adamchuk et al., 2015). The use of electrical signals at the surface (0-30 cm) or at depth (0100 cm) differs from one study to another according to several scientific and operational viewpoints in relation to the diversity of the soils studied (Johnnson et al., 2003; Peralta and Costa, 2013, Peralta et al., 2015)

New types of SPCs have been introduced in North America for the purpose of characterizing the SV of soil properties in PA, including ECa (Adamchuk and Tremblay 2017, Huang et al., 2018). Saifuzzaman et al. (2021) evaluated two types of sensors (RTK GNESS and DUALEM 21S) for the purpose of predicting SV of ECa and topography in relation to six soil properties (pH, buffer pH, organic matter (OM), P, potassium and CEC) from 12 fields in Ontario (Canada). High correlations (r ≥ 0.60) were obtained between ECa (measured using the sensors) and topography in relation to soil properties (OM, P and CEC). Of the 12 fields studied, two showed very reliable models for predicting the future spatial distribution of ECa in relation to soil OM and CEC.

1.5.2.3 Statistical methods for delimiting ZAs

There are several unsupervised methods that do not require prior information to delimit ZAs, including k-means fuzzy and ISODATA. The k-means fuzzy method is the most frequently used conventional method to delimit fields into ZAs (Schenatto et al., 2017; Servadio et al., 2017). A technique introduced by Ruspini (1969) and developed by Dunn (1973), the optimal mean fuzzy-k method is defined as a minimisation of the weighted measure of the squared distance between data points and class centroids (Cordoba et al., 2013). Fuzzy-k classification determines how similar an object is to a class by its class membership.

Several authors (Uribeetxebarria et al., 2018; Perron et al., 2018) have used the k-fuzzy method to delineate ZAs using soil ECa, or in combination with various edaphic properties such as texture (Moral et al., 2010), OM content (Kweon et al., 2012), depth, crop yield and elevation measurements (Peralta et al., 2015). Software packages that use the k-means fuzzy algorithm are FuzME and MZA. They were developed by Minasny et al. (2002) and Fridgen et al. (2004), respectively.

The ISODATA method, also known as the iterative self-organised data analysis technique, is an extension of the k-means method. This method is based on the use of a fast classification algorithm (Tou and Gonzalez, 1974) which does not require the introduction of class characteristics prior to classification (unsupervised classifier) (Guastaferro et al., 2010). However, this technique requires the variables to be normally distributed, with equal variances in order to group similar features by mean vectors and a covariance matrix (Fraisse et al., 2001). Several authors have adopted the ISODATA technique to delineate MZs (Guastaferro et al., 2010, Haghverdi et al., 2015; Abbas et al., 2016; Uribeetxebarria et al., 2018).

1.5.2.4 Determining and validating the optimum number of ZAs

The optimal number of ZAs is determined using the variance decay method (Haghverdi et al., 2015; Lajili et al., 2021). Based on this approach, the total within-area variance is expressed as a percentage of the variance for the whole field (i.e. 1 ZA) (Cambouris et al., 2006). Thus, the optimal number of ZAs is assessed using the inflection point of the variance decay curve (see Figure 1.6).

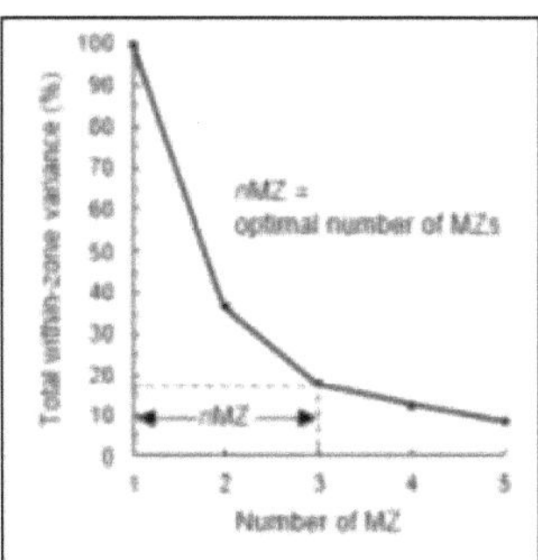

Figure 1-6. Determination of the optimal number of ZAs using the variance decay method (Adapted from Lajili et al., 2021).

After determining the optimal number of ZAs, a normality test for soil properties is performed using the SAS PROC UNIVARIATE procedure (SAS Institute, 2010). For normally distributed soil properties, an ANOVA combined with an LSD multiple comparison test (p-value < 0.05) was performed to determine whether the properties varied significantly from one ZA to another. Non-normally distributed soil properties were analysed using non-parametric tests (Wilcoxon and Kruskal-Wallis) of the NPAR1WAY Procedure (SAS Institute, 2010). In this way, significant differences in soil properties within the delimited ZAs will make it possible to: 1) determine (confirm) the extent to which the main parameter measured (i.e. the ECa) captured the intra-field spatial variability in soil properties (Peralta et al., 2013) and 2) assess the effectiveness of the delimitation of the ZAs.

Thus, it is possible to use the delineation of ZAs to control the spatial variability of soil P in PA to increase farm productivity and yields while reducing P losses to the environment. There are few examples of the application of this agronomic approach using P measurements to optimise P management in field crop soils in eastern Canada.

1.6. Assumptions and objectives

1.6.1. Assumptions

The literature review presented above showed that knowledge of P SV (available P, $[P/Al]_{M3}$) in cultivated soils according to cropping systems (young grasslands vs. old grasslands) and tillage (direct seeding vs. conventional tillage) is still very limited, particularly in arable soils. In addition, few studies have been carried out to date on the agri-environmental strategy of delimiting ZAs on the basis of P measurements for sustainable P management in field crop soils in eastern Canada. The following hypotheses and objectives were therefore put forward for this study, which is divided into three experiments:

Experiment 1

1) Soils under old grassland have a higher available P content than soils under young grassland, particularly in the 0-5 cm layer;

2) The two P availability indicators (P_{M3}, $[P/Al]_{M3}$) are strongly influenced by Al_{M3}, Fe_{M3}, Ca_{M3} and total C content in soils under old grassland compared with soils under young grassland;

3) Soils under old grassland have a lower spatial dependence on available P than soils under young grassland, particularly in the 0-5 cm layer.

Experiment 2

1) No-till farming increases the available P content in the 0-5 cm layer, compared with conventionally tilled soil;

2) The two P availability indicators (P_{M3}, $[P/Al]_{M3}$) are strongly influenced by Al_{M3}, Fe_{M3}, Ca_{M3} and total C content in no-till soils compared with conventionally tilled soils in the 0-5 cm layer;

3) No-till soils have a higher spatial dependence on available P than conventionally tilled soils, particularly in the 0-5 cm layer.

Experiment 3

1) Spatial variability of available P is high in a P-stratified soil;

2) Both P availability indicators (P_{M3}, $[P/Al]_{M3}$) are influenced by ECa in a P stratified soil;

3) The delimitation of ZAs from P measurements represents an effective strategy for reducing the spatial variability of P for the purposes of precise P2O5 recommendations in a P-stratified soil.

3.2.2. Objectives

The general objective of the thesis project is to compare the spatial and vertical variability of available P (P_{M3}, $[P/A1]_{M3}$) in cultivated soils under the effect of cropping systems (old grassland vs. young grassland) and tillage (conventional vs. no-till) at the two depths studied (0-5 cm) and (5-20 cm) for the purposes of agri-environmental recommendations for phosphate (P_{2O5}) fertilisation in PA.

The specific objectives of the thesis project are as follows:

1) To assess the spatial variability of indicators of P availability (P_{M3}, $[P/A1]_{M3}$) and other selected physico-chemical properties (pH, total carbon content, $A1_{M3}$, Ca_{M3}, Fe_{M3}) of soil under young grassland compared with soil under old grassland according to the two depths (0-5 cm) and (5-20 cm) using descriptive statistical and geostatistical tools;

2) To study the relationships between the two indicators of P availability (P_{M3}, $[P/A1]_{M3}$) and the total C, $A \mid M3$, Ca_{M3}, and Fe_{M3} contents of a soil under young grassland compared with a soil under old grassland according to the two depths (0-5 cm) and (5-20 cm);

3) To determine the spatial dependence ($C/(c+co)$) of available P (P_{M3}, $[P/A1]_{M3}$) and other elements (total C, $A \mid M3$, Fe_{M3}, and Ca_{M3}) of a soil under young grassland compared with the soil under old grassland in the two depths studied (0-5 cm) and (5-20 cm).

4) To assess the spatial variability of indicators of P availability (P_{M3}, $[P/A1]_{M3}$) and other selected physico-chemical properties (pH, total carbon content, $A \mid M3$, Ca_{M3}, Fe_{M3}) of a soil under conventional tillage compared with a soil under direct seeding according to the two depths (0-5 cm) and (5-20 cm) using descriptive statistical and geostatistical tools;

5) To study the relationships between the two indicators of P availability (P_{M3}, $[P/Al]_{M3}$) and the total C, $A \mid M3$, Ca_{M3}, and Fe_{M3} contents of a soil under conventional tillage compared with a soil under direct seeding according to the two depths (0-5 cm) and (5-20 cm);

6) To determine the spatial dependence ($C/(c+co)$) of available P (P_{M3}, $[P/Al]_{M3}$) and other elements (total C, $A \mid M3$, Fe_{M3}, and Ca_{M3}) of a soil under conventional tillage compared with a soil under direct seeding according to the two depths studied (0-5 cm) and (5-20 cm);

7) To evaluate current phosphate fertilisation recommendations using prescription maps of kriged values of $(P/A1)_{M3}$ of a soil under conventional tillage compared with a soil under direct seeding according to the two depths studied (0-5 cm) and (5-20 cm).

8) To study the variability and spatial structure of indicators of P availability (P_{M3}, $[P/A1]_{M3}$) and auxiliary properties (CE_{a30}, CE_{a100}, elevation, yields) of a no-till soil according to the depth studied (0-5 cm) using descriptive statistical and geostatistical tools;

9) To study the relationships between the two indicators of P availability (P_{M3}, $[P/Al]_{M3}$), total C, $A \mid M3$, Ca_{M3}, and Fe_{M3} in the soil and the auxiliary properties (CE_{a30}, CE_{a100}, elevation, yields) of a no-till soil according to the depth studied (0-5 cm);

10) Determine the optimum number of ZAs for a no-till soil using the variance reduction method based on the soil's CEa-P relationship at the depth studied (0-5 cm);

11) Validate the ZAs obtained for the purposes of optimising precise P fertiliser recommendations according to the depth studied (0-5 cm).

1.7 References

Abbas, A., Minalla, N., Ahmad, N., Abid, S., Khan, M. (2016). "K-Means and ISODATA clustering algorithms for landcover classification using remote sensing". Sindh University Research Journal-SURJ (Science Series) **48**(2):315-318.

Abdi, D., Cade-Menun, B., Ziadi, N., Parent, L. É. (2014). "Long-term impact of tillage practices and phosphorus fertilization on soil phosphorus forms as determined by 31 P nuclear magnetic resonance spectroscopy". Journal of Environmental Quality **43**(4):1431-1441.

Abdi, D., Cade-Menun, B., Ziadi, N., Tremblay, G. F., Parent, L. É. (2016). "Visible near infrared reflectance spectroscopy to predict soil phosphorus pools in chernozems of Saskatchewan, Canada". Geoderma Regional **7**(2): 93-101.

Adamchuk, V., Allred, B., Doolittle, J., Grote, K., Rossel, R., Ditzler, C., West, L. (2015). Tools for proximal soil sensing. Soil Survey Staff, Ditzler, C., West, L.(Eds.), Soil Survey Manual. Natural Resources Conservation Service. US Department of Agriculture Handbook, **18**.

Adamchuk, V., Tremblay, N. (2017). New developments in proximal soil sensing. In Proceedings of The International Tri-Conference for Precision Agriculture, Hamilton.

Allaire, S. E., Van Bochove, E., Denault, J.-T., Dadfar, H., Thériault, G., Charles, A., De Jong, R. (2011). "Preferential pathways of phosphorus movement from agricultural land to water bodies in the Canadian Great Lakes basin: A predictive tool". Canadian Journal of Soil Science **91**(3):361-374.

Armstrong, R., Fitzpatrick, J., Rab, M., Abuzar, M., Fisher, P., O'Leary, G. (2009). "Advances in precision agriculture in south-eastern Australia. III. Interactions between soil properties and water use help explain spatial variability of crop production in the Victorian Mallee". Crop and Pasture Science **60**(9): 870-884.

Arrouays, D., Vion, I., Jolivet, C., Guyon, D., Couturier, A. and Wilbert, J. (1997). "Intraparcel variability of some sandy soil properties in the Landes de Gascogne (France): Conséquences sur la stratégie d'echantillonnage agronomique". Étude et gestion des sols **4**:5-16.

Arvalis (2015). Precision agriculture, GPS positioning, intraparcel management, performance evaluation. Edition Arvalis. 104 pp.

Barrow, N., Shaw, T. (1980). "Effect of drying soil on the measurement of phosphate adsorption". Communications in Soil Science and Plant Analysis **11**(4): 347-353.

Beauchemin, S., Hesterberg, D., Chou, J., Beauchemin, M., Simard, R., Sayers, D. (2003). "Speciation of phosphorus in phosphorus-enriched agricultural soils using X-ray absorption near-edge structure spectroscopy and chemical fractionation". Journal of Environmental Quality **32**(5): 1809-1819.

Bechmann, M., Kleinman, P., Sharpley, A., Saporito, L. (2005). "Freeze-thaw effects on phosphorus loss in runoff from manured and catch-cropped soils". Journal of Environmental Quality **34**(6): 2301-2309.

Becker, S., Franz, T., Abimbola, O., Steele, D., Flores, J., Jia, X., Scherer, T., Rudnick, D. R., Neale, C. (2022). "Feasibility assessment on use of proximal geophysical sensors to support precision management". Vadose Zone Journal: e20228.

Beegle, D. (2005). Assessing soil phosphorus for crop production by soil testing. Phosphorus: Agriculture and the environment, 46, 123-143.

Behera, S., Shukla, A. (2015). "Spatial distribution of surface soil acidity electrical conductivity, soil organic carbon content and exchangeable potassium, calcium and magnesium in some cropped acid soils of India". Land Degradation and Development **26**: 71-79.

Bennett, E., Carpenter, S., Clayton, M. (2005). "Soil phosphorus variability: scale-dependence in an urbanizing agricultural landscape". Landscape Ecology **20**(4):389-400.

Bogrekci, I., Lee, W. (2005). "Spectral phosphorus mapping using diffuse reflectance of soils and grass". Biosystems Engineering **91**(3): 305-312.

Bogunovic, I., Mesic, M., Zgorelee, Z., Jurisic, A., Bilandzija, D. (2014). "Spatial variation of soil nutrients on sandy-loamy soil". Soil and Tillage Research **144**: 174-183.

Bowmen, R., Cole, C. (1978). "An exploratory method for fractionation of organic phosphorous from grassland". Soil Science **125**(2): 95-101.

Breeuwsma, A., Silva, S. (1992). Phosphorus fertilisation and environmental effects in the Netherlands and the Po region (Italy). Wageningen, The Netherlands: DLO The Winand Staring Centre.

Cade-Menun, B., Carter, M., James, D., Liu, C. (2010). "Phosphorus forms and chemistry in the soil profile under long-term conservation tillage: A phosphorus-31 nuclear magnetic resonance study". Journal of Environmental Quality **39**(5): 1647-1656.

Cambardella, C., Karlen, D. (1999). "Spatial analysis of soil fertility parameters". Precision Agriculture **1**: 5-14.

Cambardella, C., Moorman, T., Parkin, T., Karlen, D., Novak, J., Turco, R., Konopka, A. (1994). "Field-scale variability of soil properties in central Iowa soils". Soil Science Society of America Journal **58**(5):1501- 1511.

Cambouris, A., Messiga, A., Ziadi, N., Perron, I., Morel, C. (2017). "Decimetric-scale two-dimensional distribution of soil phosphorus after 20 years of tillage management and maintenance phosphorus fertilization". Soil Science Society of America Journal **81**(6): 1606-1614.

Cambouris, A., Nolin, M., Zebarth, B., Laverdière, M. (2006). "Soil management zones delineated by electrical conductivity to characterize spatial and temporal variations in potato yield and in soil properties". American Journal of Potato Research **83**(5):381-395.

Cambouris A., Zebarth B., Ziadi, N., Perron I. (2014). "Precision agriculture in potato production". Potato Research **57**(3-4):249-262.

Cameron, D., Nyborg, M., Toogood, J., Laverty, D. (1971). "Accuracy of field sampling for soil tests". Canadian Journal of Soil Science **51**(2):165-175.

Cathcart, J. B. (1980). World phosphate reserves and resources. The role of phosphorus in agriculture. In Khasawneh et al (ed.). American Society of Agronomy. Madison, Wisconsin, USA. p. 1-18.

Chang, C., Laird, D., Mausbach, M., Hurburgh, C. (2001). "Near-infrared reflectance spectroscopy-principal components regression analyses of soil properties". Soil Science Society of America Journal **65**(2): 480-490.

Chang, J., Clay, D., Carlson, C., Malo, D., Clay, S., Lee, J., Ellsbury, M. (1999). "Precision farming protocols: Part 1. Grid distance and soil nutrient impact on the reproducibility of spatial variability measurements". Precision Agriculture **1**(3):277-289.

Chang, S., Jackson, M. (1957). "Fractionation of soil phosphorus". Soil Science **84**, p.133-144.

Condron, L., Turner, B., Cade-Menun, B. (2005). Chemistry and dynamics of soil organic phosphorus. Phosphorus: agriculture and the environment. In: J.T. Sims and A.N. Sharpley Editors. ASA, CSSA, and SSSA, Madison, WI. 46, 87-121.

Cooperband, L., Logan, T. (1994). "Measuring in situ changes in labile soil phosphorus with anion-exchange membranes". Soil Science Society of America Journal **58**(1): 105-114.

Corbridge, D.E.C. (2000). Phosphorus: Chemistry, biogeochemistry and technology. Elsevier, New York.

Córdoba M., Bruno C., Costa J., Balzarini, M. (2013). "Subfield management class delineation using cluster analysis from spatial principal components of soil variables". Computers and Electronics in Agriculture **97**:6-14.

Corwin, D., Lesch, S. (2003). "Application of soil electrical conductivity to precision agriculture: theory, principles, and guidelines". Agronomy Journal **95**(3):455-471.

Corwin, D., Plant R. (2005). "Applications of apparent soil electrical conductivity in precision agriculture". Computers and Electronics in Agriculture **46**(1):1-10.

CRAAQ (2010). Guide de référence en fertilisation. 2ème edition. Centre de référence en agriculture et agroalimentaire du Québec, Québec. 473 pp.

Dang, Y., Moody, P., Bell, M., Seymour, N., Dalal, R., Freebairn, D., Walker, S. (2015). "Strategic tillage in notill farming systems in Australia's northern grains-growing regions: II. Implications for agronomy, soil and environment". Soil and Tillage Research **152**: 115-123.

Dick, R. (1992). "A review: long-term effects of agricultural systems on soil biochemical and microbial parameters". Agriculture, Ecosystems and Environment **40**(1-4): 25-36.

Djodjic, F., Börling, K., Bergström, L. (2004). "Phosphorus leaching in relation to soil type and soil phosphorus content". Journal of Environmental Quality **33**(2):678-684.

Doerge, T. (2011). Management zone concepts. Site-specific management guidelines. SSMG-2. Potash & Phosphate Institute. International Plant Nutrition Institute, Norcross, GA. 4 p.

Duiker, S., Beegle, D. (2006). "Soil fertility distributions in long-term no-till, chisel/disk and moldboard plow/disk systems". Soil and Tillage Research **88**(1-2): 30-41.

Dunn, J. (1973). "A fuzzy relative of the ISODATA process and its use in detecting compact well-Separated clusters". Journal of Cybernetics **3**:3, 32-57.

Fageria, N., He, Z., Baligar, V. (2017). Phosphorus Management in Crop Production. Taylor and Francis Group, CRC Press. 360 pp.

Ferguson, R., Hergert, G. (2009). Soil sampling for precision agriculture. University of Nebraska Lincoln, Extension EC-154.

Ferguson, R., Hergert, G., Schepers, J., Gotway, C., Cahoon, J., Peterson, T. (2002). "Site-specific nitrogen management of irrigated maize: yield and soil residual nitrate effects". Soil Science Society of America Journal **66**: 544-553.

Fife, C. (1963). "An evaluation of ammonium fluoride as a selective extractant for aluminum-bound soil phosphate: detailed study of soils". Soil Science **96**(2): 112-120.

Fraisse, C., Sudduth K., Kitchen, N. (2001). "Delineation of site-specific management zones by unsupervised classification of topographic attributes and soil electrical conductivity". Transactions of the ASAE **44**(1):155-166.

Franzen, D. (2018). Soil variability and fertility management. In book Precision agriculture basics, published by American Society of Agronomy, Crop Science Society of America, and Soil Science Society of America. Pages 79-92.

Freppaz, M., Williams, B., Edwards, A., Scalenghe, R., Zanini, E. (2007). "Simulating soil freeze/thaw cycles typical of winter alpine conditions: implications for N and P availability". Applied Soil Ecology **35**(1): 247-255.

Fridgen J., Kitchen, N., Sudduth, K. (2000). Variability of soil and landscape attributes within subfield management zones. In Proceedings of the 5th International Conference on Precision Agriculture, Bloomington, Minnesota, USA, 16-19 July, 2000. American Society of Agronomy, p 1-16.

Fridgen J., Kitchen N., Sudduth K., Drummond S., Wiebold W., Fraisse C. (2004). "Management zone analyst (MZA) software for subfield management zone delineation". Agronomy Journal **96**(1):100-108.

Friedrich, T., Derpsch, R., Kassam, A. (2012). Overview of the global spread of conservation agriculture. Field Actions Science Reports. The journal of field actions, (Special Issue 6).

Frossard, E., Julien, P., Neyroud, J., Sinaj, S. (2004). Phosphorus in soils: the situation in Switzerland: phosphorus in soils, fertilisers, crops and the environment. Swiss Agency for the Environment, Forests and Landscape SAEFL.

Fu, W., Zhao, K., Jiang, P., Ye, Z., Tunney, H., Zhang, C. (2013). "Field-scale variability of soil test phosphorus and other nutrients in grasslands under long-term agricultural managements". Soil Research **51**(6):503-512.

Gagnon, B., Ziadi, N., Côté, C., Foisy, M. (2010). "Environmental impact of repeated applications of combined paper mill biosolids in silage corn production". Canadian Journal of Soil Science **90**(1):215-227.

Gassner, A. (2003). Factors controlling the spatial specification of phosphorus in agricultural soils. Special Issue 244, Federal Agricultural Research Center, Braunschweig, Germany.

Gburek, W., Sharpley, A., Heathwaite, L., Folmar, G. (2000) "Phosphorus management at the watershed scale: a modification of the phosphorus index". Journal of Environmental Quality **29**(1):130-144.

Gilbert, N. (2009). "Environment: the disappearing nutrient. Nature **461**:716-718.

Giroux, M., Cantin, J., Rivest, R. and Tremblay, G. (2002). Changes in phosphorus levels in soils according to fertility, phosphorus richness and soil type. Proc. Colloque sur le phosphore OAQ-APAQ-Une gestion éclairée.

Giroux, M., Duchemin, M., Michaud, A., Beaudin, I., Landry, C., Enright, P., Madramootoo, C. and Laverdière, M. (2008). "Relationship between particulate and dissolved phosphorus concentrations in runoff water and total and assimilable phosphorus levels in soils for different crops". Agrosolutions **19**:4-14.

Giroux, M. and Tran, T. (1985). "Assessment of assimilable phosphorus in acid soils with different extraction methods in relation to oat yield and soil properties". Canadian Journal of Soil Science **65**(1): 47-60.

Giroux, M. and Tran, T. (1996). "Agronomic and environmental criteria related to the availability, solubility and saturation of phosphorus in agricultural soils in Quebec. Agrosol **9**(2): 51-57.

Grant, C., Bittman, S., Montreal, M., Plenchette, C., Morel, C. (2005). "Soil and fertilizer phosphorus: Effects on plant P supply and mycorrhizal development". Canadian Journal of Plant Science **85**(1): 3-14.

Grant, R., Laubel, A., Kronvang, B., Andersen, H., Svendsen, L., Fuglsang, A. (1996). "Loss of dissolved and particulate phosphorus from arable catchments by subsurface drainage". Water Research **30**(11):2633-2642.

Guastaferro, F., Castrignanò, A., De Benedetto, D., Sollitto, D., Troccoli, A., Cafarelli, B. (2010). "A comparison of different algorithms for the delineation of management zones". Precision Agriculture **11**(6):600-620.

Haghverdi, A., Leib, B., Washington-Allen, R., Ayers, P., Buschermohle, M. (2015). "Perspectives on delineating management zones for variable rate irrigation". Computers and Electronics in Agriculture **117**:154-167.

Haneklaus, S., Schnug, E. (2006). Site Specific Nutrient Management: Objectives, Current Status, and Future Research Needs. In Handbook of Precision Agriculture: Principles and Applications, 91. CRC Press.

Hébert, M., Busset, G., Groeneveld, E. (2008). Bilan 2007 de la valorisation des matières résiduelles fertilisantes. Ministère du Développement durable, de l'Environnement et des Parcs. Québec,14 pages.

Hedley, M., Stewart, J., Chauhan, B. (1982). "Changes in inorganic and organic soil phosphorus fractions induced by cultivation practices and by laboratory incubations". Soil Science Society of America Journal **46**(5): 970-976.

Heilig, J., Kempenich, J., Doolittle, J., Brevik, E., Ulmer, M. (2011). "Evaluation of electromagnetic induction to characterize and map sodium-affected soils in the Northern Great Plains". Soil Survey Horizons **52**(3):77-88.

Heiniger, R., McBride, R., Clay, D. E. (2003). "Using soil electrical conductivity to improve nutrient management". Agronomy Journal **95**(3): 508-519.

Hinsinger, P. (1998). "How do plant roots acquire mineral nutrients? Chemical processes involved in the rhizosphere". Advances in Agronomy **64**: 225-265.

Hinsinger, P. (2001). "Bioavailability of soil inorganic P in the rhizosphere as affected by root-induced chemical changes: a review". Plant and Soil **237**(2): 173-195.

Hinsinger, P., Gobran, G., Gregory, P., Wenzel, W. (2005). "Rhizosphere geometry and heterogeneity arising from root-mediated physical and chemical processes". New Phytologist **168**(2): 293-303.

Hinsinger, P., Jaillard, B., Le Cadre, E. and Plassard, C. (2007). "Phosphorus speciation and bioavailability in the rhizosphere". Océanis **33-1/2**: 37-50.

Holm, F., Zentner, R., Thomas, A., Sapsford, K., Légère, A., Gossen, B., ..., Leeson, J. (2006). "Agronomic and economic responses to integrated weed management systems and fungicide in a wheat-canola-barley- pea rotation". Canadian Journal of Plant Science **86**(4): 1281-1295.

Huang, H., Adamchuk, V., Biswas, A., Ji, W., Lauzon, S. (2018). Analysis of soil properties predictability using different on-the-go soil mapping systems. In Proceedings of the 14th International Conference on Precision Agriculture. Montreal, Quebec, Canada. p. 24-27.

Ige, D., Akinremi, O., Flaten, D. (2005). "Environmental index for estimating the risk of phosphorus loss in calcareous soils of Manitoba". Journal of Environmental Quality **34**(6): 1944-1951.

Jamieson, A., Madramootoo, C., Enright, P. (2003). "Phosphorus losses in surface and subsurface runoff from a snowmelt event on an agricultural field in Quebec". Canadian Biosystems Engineering **45**:1-1.

Jia, S.; Zhou, D., Xu, D. (2011). "The temporal and spatial variability of soil properties in an agricultural system as affected by farming practices in the past 25 years". Journal of Food, Agriculture and Environment **9**: 669-676.

Johnson, C., Mortensen, D., Wienhold, B., Shanahan, J., Doran, J. (2003). "Site-specific management zones based on soil electrical conductivity in a semiarid cropping system". Agronomy Journal **95**(2):303-315.

Jordan, C., McGuckin, S., Smith, R. (2000) "Increased predicted losses of phosphorus to surface waters from soils with high Olsen-P concentrations". Soil Use and Management **16**(1):27-35.

Khiari, L., Parent, L. É., Pellerin, A., Alimi, A., Tremblay, C., Simard, R., Fortin, J. (2000) "An agri-environmental phosphorus saturation index for acid coarse-textured soils". American Society of Agronomy, Crop Science Society of America, and Soil Science Society of America **29** (5):1561-1567.

Khosla, R., Alley, M. (1999). "Soil-specific nitrogen management on mid-Atlantic coastal plain soils". Better Crops **83**(3): 6-7.

Khosla, R., Fleming, K., Delgado, J., Shaver, T., Westfall, D. (2002). "Use of site-specific management zones to improve nitrogen management for precision agriculture". Journal of Soil and Water Conservation **57**(6): 513-518.

Kitchen, N., Clay, S. (2018). Understanding and identifying variability. In book Precision agriculture basics, published by American Society of Agronomy, Crop Science Society of America, and Soil Science Society of America. Pages 13-24.

Kitchen, N., Sudduth, K., Myers, D., Drummond, S., Hong, S. (2005). "Delineating productivity zones on claypan soil fields using apparent soil electrical conductivity". Computers and Electronics in Agriculture **46**(1):285-308.

Klatt, J., Mallarino, A., Downing, J., Kopaska, J., Wittry, D. (2003). "Soil phosphorus, management practices, and their relationship to phosphorus delivery in the Iowa Clear Lake agricultural watershed". Journal of Environmental Quality **32**(6):2140-2149.

Konstantinos, K., Apostolos, X., Panagiotis, K., George, S. (2007). Topology optimization in wireless sensor networks for precision agriculture applications. In International Conference on Sensor Technologies and

Applications (SensorComm 2007). IEEE, p 526-530.

Kovar, J., Barber, S. (1988). "Phosphorus supply characteristics of 33 soils as influenced by seven rates of phosphorus addition". Soil Science Society of America Journal **52**(1): 160-165.

Kuo, S. (1996). Phosphorus. In Methods of soil analysis. Part 3, Chemical Methods. Sparks, D.L. (eds). SSSA book series: 5. pp: 869-919.

Kweon, G. (2012). "Delineation of site-specific productivity zones using soil properties and topographic attributes with a fuzzy logic system". Biosystems Engineering **112**(4):261-277.

Lafond, G., Walley, F., May, W., Holzapfel, C. (2011). "Long term impact of no-till on soil properties and crop productivity on the Canadian prairies". Soil and Tillage Research **117**: 110-123.

Lajili, A., Cambouris, A. N., Chokmani, K., Duchemin, M., Perron, I., Zebarth, B., Biswas, A., Adamchuk, V. (2021). "Analysis of four delineation methods to identify potential management zones in a commercial potato field in Eastern Canada". Agronomy **11**(3): 432.

Li, H. (2017). Long-term impact of tillage on the biogeochemical cycle of phosphorus: analysis of the L'Acadie trial (Quebec, Canada) and modelling. Joint PhD thesis in Soils and Environment, Quebec, Université Laval and France, Université de Bordeaux. 197 pages.

Li, Y., Shi, Z., Li, F., Li, H. (2007). "Delineation of site-specific management zones using fuzzy clustering analysis in a coastal saline land". Computers and Electronics in Agriculture **56**(2): 174-186.

Lund, E., Colin, P., Christy, D., Drummond, P. (1999). Applying soil electrical conductivity technology to precision agriculture. In Proceedings of the fourth international conference on precision agriculture. American Society of Agronomy, Crop Science Society of America, Soil Science Society of America. Madison, WI, USA. Pages 1089-1100.

Mallarino, A. (1996). "Spatial variability patterns of phosphorus and potassium in no-tilled soils for two sampling scales". Soil Science Society of America Journal **60**(5): 1473-1481.

Mallarino, A., Beegle, D., Joern, B. (2007). Soil sampling methods for phosphorus. Minimizing phosphorus losses from agriculture.

Mallarino, A., Beegle D., Joern, B. (2014). Soil Sampling Methods for Phosphorus - Spatial concerns. A SERA-17 Position Paper.

Malley, D., Martin, P., Ben-Dor, E. (2004). Application in analysis of soils. In C. A. Roberts, J. Workman Jr, and J. B. Reeves III, eds. Agronomy 44. ASA, CSSA, SSSA, Madison, WI. Near-Infrared Spectroscopy in Agriculture **44**: 729-784.

Mason, S., McLaughlin, M., Johnston, C., McNeill, A. (2013). "Soil test measures of available P (Colwell, resin and DGT) compared with plant P uptake using isotope dilution". Plant and Soil **373**(1): 711-722.

McBratney, A., Pringle, M. (1999). "Estimating average and proportional variograms of soil properties and their potential use in precision agriculture". Precision Agriculture **1**(2):125-152.

McGechan, M. (2002). "Sorption of phosphorus by soil, part 2: measurement methods, results and model parameter values". Biosystems Engineering **82**(2): 115-130.

McGechan, M., Lewis, D. (2002). "Sorption of phosphorus by soil, part 1: principles, equations and models". Biosystems Engineering **82**(1): 1-24.

MDDELCC. (2017). Guide de Référence du Règlement sur les Exploitations Agricoles. Ministère du Développement durable, de l'Environnement et de la Lutte contre les changements climatiques: Québec, QC, Canada; p. 185.

Mertens, F., Pätzold, S., Welp, G. (2008). "Spatial heterogeneity of soil properties and its mapping with apparent electrical conductivity". Journal of Plant Nutrition and Soil Science **171**(2):146-154.

Messiga, A. (2010). Phosphorus transfers in field crop soils. Doctoral thesis in Soils and Environment, Québec, Université Laval. 217 pages.

Messiga, A., Ziadi, N., Morel, C., Parent, L. É. (2010). "Soil phosphorus availability in no-till versus conventional tillage following freezing and thawing cycles". Canadian Journal of Soil Science **90**(3): 419-428.

Messiga, A., Ziadi, N., Morel, C., Grant, C., Tremblay, G., Lamarre, G., Parent, L. É. (2012). "Long term impact of tillage practices and biennial P and N fertilization on maize and soybean yields and soil P status". Field Crops Research **133**: 10-22.

Minasny, B, McBratney, A. (2002). FuzME version 3.0. Australian Centre for Precision Agriculture. The University of Sydney, Australia.

Mondal, B., Sekhon, S., Sadhukhan, R., Singh, R. (2020). "Spatial variability assessment of soil available phosphorus using geostatistical approach". Indian Journal of Agricultural Sciences **90** (6): 1170-1175.

Moral, F., Terrón, J., Da Silva, J. (2010). "Delineation of management zones using mobile measurements of soil apparent electrical conductivity and multivariate geostatistical techniques". Soil and Tillage Research **106**(2):335-343.

Morel, C. (2002). Characterisation of the phyto-availability of soil phosphorus by modelling the transfer of phosphate ions between soil and solution. Habilitation à diriger des Recherches. INPL-ENSA-IA Nancy. 80 pp.

Morel, C. (2007). "Mobility and bioavailability of phosphorus in cultivated soils: mechanisms, modelling and diagnosis". Oceanis **33**(1-2):51-74.

Mulla, D. (1989). Soil spatial variability and methods of analysis. In: Soil, crop, and water management systems for rainfed agriculture in the Sudano-Sahelian Zone: Proceedings of an International Workshop, 11-16 Jan. 1987. ICRISAT Sahelian Center, Niamey, Niger. pp. 241-252.

Nadler, A., Frenkel, H. (1980). "Determination of soil solution electrical conductivity from bulk soil electrical conductivity measurements by the four-electrode method". Soil Science Society of America Journal **44**(6):1216-1221.

Nduwamungu, C., Ziadi, N., Parent, L. É., Tremblay, G. (2009a). "Mehlich 3 extractable nutrients as determined by near-infrared reflectance spectroscopy". Canadian Journal of Soil Science **89**(5): 579-587.

Nduwamungu, C., Ziadi, N., Parent, L. É., Tremblay, G. F., Thuriès, L. (2009b). "Opportunities for, and limitations of, near infrared reflectance spectroscopy applications in soil analysis: A review". Canadian Journal of Soil Science **89**(5): 531-541.

Nolan, S., Little, J., Casson, J., Hecker, F., Olson, B. (2007). "Field-scale variation of soil phosphorus within small Alberta watersheds". Journal of Soil and Water Conservation **62**(6):414-422.

Nolin, M. and Caillier, M. (1992a). "Soil variability. I-Components and causes". Agrosol **5**(1):15-20.

Nolin, M. and Caillier, M. (1992b). "La variabilité des sols. II-Quantification et amplitude". Agrosol **5**(1):21-32.

Nyiraneza, J., Nolin, M., Ziadi, N., Cambouris, A. (2011). "Short-range variability of nitrate and phosphate desorbed from anionic exchange membranes". Soil Science Society of America Journal **75**(6): 22422250.

O'Halloran, I. (1993). "Effect of tillage and fertilization on inorganic and organic soil phosphorus". Canadian Journal of Soil Science **73**(3): 359-369.

O'Halloran, I., Cade-Menun, B. (2007). Total and organic phosphorus. In Soil Sampling and Methods of analysis. Carter, M.R., and Gregorich, E.G., (2nd eds). p. 265-291.

Olsen, S. (1954). Estimation of available phosphorus in soils by extraction with sodium bicarbonate (No. 939). US Department of Agriculture.

Olson, K., Ebelhar, S. (2009). "Impacts of conservation tillage systems on long-term crop yields". Journal of Agronomy **8**(1): 14-20.

Page, T., Haygarth, P., Beven, K., Joynes, A., Butler, T., Keeler, C., Freer, J., Owens, P. N., Wood, G. (2005). "Spatial variability of soil phosphorus in relation to the topographic index and critical source areas: sampling for assessing risk to water quality". Journal of Environmental Quality **34**(6):2263-2277.

Pellerin, A., Parent, L. É., Tremblay, C., Fortin, J., Tremblay, G., Landry, C., Khiari, L. (2006). "Agri-environmental models using Mehlich-III soil phosphorus saturation index for corn in Quebec". Canadian Journal of Soil Science **86**(5): 897-910.

Peralta, N., Costa, J. (2013). "Delineation of management zones with soil apparent electrical conductivity to improve nutrient management". Computers and Electronics in Agriculture **99**:218-226.

Peralta, N., Costa, J., Balzarini, M., Franco, M., Córdoba, M., Bullock, D. (2015). "Delineation of management zones to improve nitrogen management of wheat". Computers and Electronics in Agriculture **110**:103113.

Perron, I., Cambouris, A. N., Chokmani, K., Vargas Gutierrez, M., Zebarth, B., Moreau, G., Biswas, A., Adamchuk, V. (2018). "Delineating soil management zones using a proximal soil sensing system in two commercial potato fields in New Brunswick, Canada". Canadian Journal of Soil Science **98**(4): 724737.

Petersen, G., Corey, R. (1966). "A modified Chang and Jackson procedure for routine fractionation of inorganic soil phosphates". Soil Science Society of America Journal **30**(5): 563-565.

Piegholdt, C., Geisseler, D., Koch, H., Ludwig, B. (2013). "Long-term tillage effects on the distribution of phosphorus fractions of loess soils in Germany". Journal of Plant Nutrition and Soil Science **176**(2): 217-226.

Pierzynski, G., McDowell, R., Thomas Sims, J. (2005). Chemistry, cycling, and potential movement of inorganic phosphorus in soils. Phosphorus: Agriculture and the environment. In Agronomy monograph No. 46. Sims, J.T., and Sharpley A.N. (Eds). p 51-86.

Pittelkow, C., Liang, X., Linquist, B., Van Groenigen, K., Lee, J., Lundy, M., ..., Van Kessel, C. (2015). "Productivity limits and potentials of the principles of conservation agriculture". Nature **517**(7534): 365368.

Puustinen, M., Koskiaho, J., Peltonen, K. (2005). "Influence of cultivation methods on suspended solids and phosphorus concentrations in surface runoff on clayey sloped fields in boreal climate". Agriculture, Ecosystems and Environment **105**(4): 565-579.

Qian, P., Schoenau, J., Huang, W. (1992). "Use of ion exchange membranes in routine soil testing". Communications in Soil Science and Plant Analysis **23**(15-16): 1791-1804.

Quenum, M., Nolin, M., Bernier, M. (2012). "Digital mapping of the maximum phosphorus sorption capacity of soils at the agricultural plot scale using auxiliary variables". Canadian Journal of Soil Science **92**(5): 733-750.

Rab, M., Fisher, P., Armstrong, R., Abuzar, M., Robinson, N., Chandra, S. (2009). "Advances in precision agriculture in south-eastern Australia. IV. Spatial variability in plant-available water capacity of soil and its relationship with yield in site-specific management zones". Crop and Pasture Science **60**(9): 885900.

Redding, M., Shatte, T., Bell, K. (2006). "Soil sorption-desorption of phosphorus from piggery effluent compared with inorganic sources". European Journal of Soil Science **57**(2): 134-146.

Rhoades, J., Manteghi, N., Shouse, P., Alves, W. (1989). "Soil electrical conductivity and soil salinity: new formulations and calibrations". Soil Science Society of America Journal **53**(2): 433-439.

Rivero, E., Cruzate, G., Beltrán, M., Russo, S., Mallarino, A. (2012). "Spatial variability of phosphorus and its relationship with some properties of soils in Argentina Republic". Journal of Environmental Science and Engineering **A1**: 797-801.

Robinson, T., Mettermicht, G. (2006). "Testing the performance of spatial interpolation techniques for mapping soil properties". Computer and Electronics in Agriculture **50**(2): 97-108.

Rodrigués, M., Pavinato, P., Withers, P., Teles, A., Herrera, W. (2016). "Legacy phosphorus and no tillage agriculture in tropical oxisols of the Brazilian savanna". Science of the Total Environment **542**: 10501061.

Roger, A., Libohova, Z., Rossier, N., Joost, S., Maltas, A., Frossard, E., Sinaj, S. (2014). "Spatial variability of soil phosphorus in the Fribourg canton, Switzerland". Geoderma **217**:26-36.

Ruspini, E. (1969). "A new approach to clustering". Information and Control **15**(1):22-32.

Saglam, M., Ozturk, H., Ersahin, S., Ozkan, A. (2011). "Spatial variation of soil physical properties in adjacent alluvial and colluvial soils under ustic moisture regime". Hydrology and Earth System Sciences Discussions **8** (2): 4261-4280.

Saifuzzaman, M., Adamchuk, V., Biswas, A., Rabe, N. (2021). "High-density proximal soil sensing data and topographic derivatives to characterise field variability". Biosystems Engineering **211**: 19-34.

SAS Institute. (2010). SAS user's guide. Statistics. Version 9.3. SAS Inst., Cary, NC. USA. SAS Institute.

Sato, S., Comerford, N. (2006). "Assessing methods for developing phosphorus desorption isotherms from soils using anion exchange membranes". Plant and Soil **279**(1): 107-117.

Schenatto, K., De Souza, E., Bazzi, C., Gavioli, A., Betzek, N., Beneduzzi, H. (2017). "Normalization of data for delineating management zones". Computers and Electronics in Agriculture **143**:238-248.

Selles, F., McConkey, B., Campbell, C. (1999). "Distribution and forms of P under cultivator-and zero-tillage for continuous-and fallow-wheat cropping systems in the semi-arid Canadian prairies". Soil and Tillage Research **51**(1-2): 47-59.

Servadio, P., Bergonzoli, S., Verotti, M. (2017). "Delineation of management zones based on soil mechanicalchemical properties to apply variable rates of inputs throughout a field (VRA)". Engineering in Agriculture, Environment and Food **10**(1):20-30.

Shainberg, I., Rhoades, J., Prather, R. (1980). "Effect of exchangeable sodium percentage, cation exchange capacity, and soil solution concentration on soil electrical conductivity". Soil Science Society of America Journal

44(3):469-473.

Shi, Y., Ziadi, N., Messiga, A., Lalande, R., Hu, Z. (2013). "Changes in soil phosphorus fractions for a long-term corn-soybean rotation with tillage and phosphorus fertilization". Soil Science Society of America Journal **77**(4): 1402-1412.

Shober, A., Hesterberg, D., Sims, J., Gardner, S. (2006). "Characterization of phosphorus species in biosolids and manures using XANES spectroscopy". Journal of Environmental Quality **35**(6): 1983-1993.

Si, B., Kachanoski, R., Reynolds, W. (2007). Analysis of soil variability. In Soil Sampling and Methods of Analysis. Second Edition. Published by Canadian Society of Soil Science. Taylor & Francis, Boca Raton, FL. Pages 1163-1191.

Simard, R., Beauchemin, S., Haygarth, P. (2000) "Potential for preferential pathways of phosphorus transport". Journal of Environmental Quality **29**(1):97-105.

Sims, J., Maguire, R., Leytem, A., Gartley, K., Pautler, M. (2002). "Evaluation of Mehlich-3 as an agri-environmental soil phosphorus test for the Mid-Atlantic United States of America". Soil Science Society of America Journal **66**(6): 2016-2032.

Smillie, G., Syers, J. (1972). "Calcium fluoride formation during extraction of calcareous soils with fluoride: II. Implications to the Bray P-1 test". Soil Science Society of America Journal **36**(1): 25-30.

Smith, S.E., Read, D.J. (1997). Mycorrhizal symbiosis. 2nd ed. Academic Press, San Diego, C.A, USA.

Statistics Canada (2016). 2011 farm and farm operator data. http://www.statcan.gc.ca/pub/95-640-x/2011001/p1/p1-05-eng.htm#XV

Steen, I. (1998). "Phosphorus availability in the 21st century: management of a non-renewable resource". Phosphorus and Potassium (**217**): 25-31.

Stevenson, F. (1994). Humus chemistry: genesis, composition, reactions. John Wiley & Sons.

Sudduth, K., Kitchen, N., Bollero, G., Bullock, D., Wiebold, W. (2003). "Comparison of electromagnetic induction and direct sensing of soil electrical conductivity". Agronomy Journal **95**(3):472-482.

Sudduth, K., Kitchen, N., Wiebold, W., Batchelor, W., Bollero, G., Bullock, D., Clay, D., Palm, H., Pierce, F., Schuler, R., Thelen, K. (2005). "Relating apparent electrical conductivity to soil properties across the north-central USA". Computers and Electronics in Agriculture **46**(1-3):263-283.

Sun, W., Huang, B., Qu, M., Tian, K., Yao, L., Fu, M., Yin, L. (2015). "Effect of farming practices on the variability of phosphorus status in intensively managed soils". Pedosphere **25**: 438-449.

Tiessen, H., Moir, J. (1993). Characterization of available P by sequential extraction. In Soil Sampling and Methods of analysis. Ed Carter, MR. 7, p. 5-229.

Tiessen, H., Moir, J. (2007). Characterization of available P by sequential extraction. In Soil Sampling and Methods of analysis. Carter, M.R., and Gregorich, E.G., (2nd eds). p. 293-306.

Tiessen, H., Stewart, J., Bettany, J. (1982). "Cultivation effects on the amounts and concentration of carbon, nitrogen, and phosphorus in grassland soils". Agronomy Journal **74**(5): 831-835.

Tou, J., González, R. (1974). Pattern recognition principles. Applied Mathematics and Computation. Reading (MA): Addison-Wesley.

Tran, T. and Giroux, M. (1985). "Comparison of different methods of extracting assimilable phosphorus in relation to the chemical and physical properties of Quebec soils". Canadian Journal of Soil Science **65**(1): 35-46.

Tran, T. and Giroux, M. (1990). "Relationship between soil properties and phosphorus availability to the plant". Agrosol **3**(1): 1.

Tran, T., Giroux, M., Fardeau, J. (1988). "Effects of soil properties on plant-available phosphorus determined by the isotopic dilution phosphorus-32 method". Soil Science Society of America Journal **52**(5):1383- 1390.

Tran, T., Giroux, M., Guilbeault, J. and Audesse, P. (1990). "Evaluation of Mehlich-III extractant to estimate the available P in Quebec soils". Communications in Soil Science and Plant Analysis **21**(1-2): 1-28.

Tunney, H. (1990). "A note on a balance sheet approach to estimating the phosphorus fertiliser needs of agriculture". Irish Journal of Agricultural Research **29**(2):149-154.

Turner, B., Haygarth, P. (2001). "Phosphorus solubilization in rewetted soils. Nature **411**(6835): 258-258.

Turner, B., Cade-Menun, B., Condron, L., Newman, S. (2005). "Extraction of soil organic phosphorus". Talanta **66**(2): 294-306.

Uribeetxebarria, A., Arnó, J., Martínez-Casasnovas, J. (2018). "Apparent electrical conductivity and multivariate analysis of soil properties to assess soil constraints in orchards affected by previous parcelling". Geoderma **319**:185-193.

US Geological Survey (2011). Mineral Commodity Summaries 2011. U.S. Geological Survey. 198 pages.

US Geological Survey (2022). Mineral Commodity Summaries 2022. U.S. Geological Survey. 202 pages. https://pubs.usgs.gov/periodicals/mcs2022/mcs2022.pdf .

Van Groenigen, J., Mutters, C., Horwath, W., Van Kessel, C. (2003). "NIR and DRIFT-MIR spectrometry of soils for predicting soil and crop parameters in a flooded field". Plant and Soil **250**(1): 155-165.

Vasu, D., Singh, S., Sahu, N., Tiwary, P., Chandran, P., Duraisami, V., Ramamurthy, V., Lalitha, M., Kalaiselvi, B. (2017). "Assessment of spatial variability of soil properties using geospatial techniques for farm level nutrient management". Soil and Tillage Research **169**:25-34.

Vrindts, E., Mouazen, A., Reyniers, M., Maertens, K., Maleki, M. R., Ramon, H., De Baerdemaeker, J. (2005). "Management zones based on correlation between soil compaction, yield and crop data". Biosystems Engineering **92**(4): 419-428.

Wagenet, R. J. (1985). Measurement and interpretation of spatially variable leaching processes. In Soil Spatial Variability. Centre for Agricultural Publishing and Documentation (PUDOC). Editors D R Nielsen and J Bouma.Wageningen, The Netherlands. pp 209-230.

Webb, J., Mallarino, A., Blackmer, A. (1992). "Effects of residual and annually applied phosphorus on soil test values and yields of corn and soybean". Journal of Production Agriculture **5**(1):148-152.

Whelan, B., McBratney, A. (2000) "The "null hypothesis" of precision agriculture management". Precision Agriculture **2**(3):265-279.

Williams, B., Hoey, D. (1987). "The use of electromagnetic induction to detect the spatial variability of the salt and clay contents of soils". Soil Research **25**(1):21-27.

Williams, J., Syers, J., Walker, T. (1967). "Fractionation of soil Inorganic phosphate by a modification of Chang and Jackson's procedure". Soil Science Society of America Journal **31**(6): 736-739.

Wilson, H., Satchithanantham, S., Moulin, A., Glenn, A. (2016). "Soil phosphorus spatial variability due to landform, tillage, and input management: A case study of small watersheds in southwestern Manitoba". Geoderma **280**:14-21.

Y   ang, J., Skogley, E., Georgitis, S., Schaff, B., Ferguson, A. (1991). "Phytoavailability soil test: development and verification of theory". Soil Science Society of America Journal **55**(5): 1358-1365.

Y   ao, R., Yang, J., Zhang, T., Gao, P., Wang, X., Hong, L., Wang, M. (2014). "Determination of site-specific management zones using soil physico-chemical properties and crop yields in coastal reclaimed farmland". Geoderma **232**: 381-393.

Y   uan, Y., Locke, M., Gaston, L. A. (2009). "Tillage effects on soil properties and spatial variability in two Mississippi Delta watersheds". Soil Science **174**(7):385-394.

Zebarth, B., Drury, C., Tremblay, N., Cambouris, A. (2009). "Opportunities for improved fertilizer nitrogen management in production of arable crops in eastern Canada: a review". Canadian Journal of Soil Science **89** (2):113-132.

Zehetner, F., Wuenscher, R., Peticzka, R., Unterfrauner, H. (2018). "Correlation of extractable soil phosphorus (P) with plant P uptake: 14 extraction methods applied to 50 agricultural soils from Central Europe". Plant, Soil and Environment **64**(4): 192-201.

Zhang, P., Wang, Y., Zhang, X. (2021). "Effects of the sampling spacing on the spatial variability in soil organic carbon, total nitrogen, and total phosphorus across a semiarid watershed". Archives of Agronomy and Soil Science **67**(10): 1359-1374.

Zheng, Z., Zhang, T. (2011). Soil phosphorus tests and transformation analysis to quantify plant availability: a review. Soil fertility improvement and integrated nutrient management: a global perspective 1, 19-36.

Zheng, Z., Simard, R., Parent, L. É. (2003). "Anion exchange and Mehlich-III phosphorus in Humaquepts varying in clay content". Soil Science Society of America Journal **67**(4): 1287-1295.

Ziadi, N., Simard, R., Allard, G., Parent, G. (2000) "Yield response of forage grasses to N fertilizer as related to spring soil nitrate sorbed on anionic exchange membranes". Canadian Journal of Soil Science **80**(1): 203-212.

Ziadi, N., Whalen, J., Messiga, A., Morel, C. (2013). "Assessment and modeling of soil available phosphorus in sustainable cropping systems". Advances in Agronomy **122**: 85-126.

Chapter 2

Spatial Variability of Soil Phosphorus Indices under Two Contrasting Grassland Fields in Eastern Canada

Jeff D. Nze Memiaghe[1,2] , Athyna N. Cambouris[1,] , Noura Ziadi[1] , Antoine Karam[2] and Isabelle Perron[1]*

[1]Agriculture and Agri-Food Canada, Quebec Research and Development Centre, 2560 Hochelaga Boulevard, Québec City, QC G1V 2J3, Canada;

[2]Soils and Agri-Food Engineering Department, Université Laval, 2425 rue de l'Agriculture, Québec City, QC G1V 0A6,

Published in Agronomy MDPI2021, 11, 24. ***https://doi.org/10.3390/agronomy11010024***

2.1. Core Ideas

- A soil P accumulation increased the potential agri-environmental risk for P pollution in permanent grasslands.

- Repeated applications of organic fertilizers increased soil P buildup in permanent grasslands while decreasing soil P variability and spatial dependence.

- A soil sampling strategy focusing on the 0-5 cm layer should be retained in permanent grasslands for sustainable P recommendations comparatively to the current soil sampling strategy (0-17.5 cm) used in Eastern Canada.

Keywords: P extracted Mehlich-3; $(P/Al)_{M3}$ index; organic manure; geostatistics; spatial distribution maps; permanent grasslands; precision agriculture.

2.2. Summary

Phosphorus (P) is an essential nutrient for growing grassland. However, continuous applications of P fertilisers lead to accumulations of P in the soil, increasing the risk of loss of this element through runoff and erosion. The overall aim of this study was to investigate the spatial variability of P indicators in two commercial fields with contrasting grassland systems using descriptive statistics and geostatistical tools in order to develop precise recommendations on soil sampling strategy and agronomic approaches related to sustainable P management in a precision farming context. Two soils, one under young grassland (JP; 2.4 ha; 2 years) and one under old grassland (AP; 2.5 ha; 10 years under permanent pasture), classified as Humo-Ferric Podzols, received organic amendments (manure, cattle slurry). A triangular grid measuring 16 m by 16 m was used to sample the soils at two depths (0-5 and 5-20 cm). Available P and other soil elements were analysed using the Mehlich-3 (M3) method. The agri-environmental indicator of P saturation $(P/Al)_{M3}$ was calculated. An accumulation of P was observed in the AP layer (0-5 cm) following long-term manure applications. Repeated applications of organic amendments may have a long-term impact on soil P accumulation, reducing the variability and spatial dependence of P in permanent grasslands. A soil sampling strategy focusing on the 0-5 cm layer should be retained in permanent grasslands for phosphate fertilizer recommendations for eastern Canadian soils.

2.3. Abstract

Phosphorus (P) is an essential nutrient for grassland production systems. However, continuous applications of P fertilizers result in soil P accumulations, increasing the risk of P losses in runoff and erosion. This study aims to investigate the field-scale variability of soil-test P (STP) in two contrasting grassland fields using descriptive statistics and geostatistics for accurate recommendations on soil sampling strategy and sustainable approaches to P management. A young grassland (YG; 2 years) and an old grassland (OG; 10 years under permanent pasture) were classified as humo-ferric podzol and received organic fertilizers. Soil samples were collected in 16-m by 16-m triangular grids at two depths (0-5 and 5-20 cm). They were analysed for available P and other soil elements extracted using the Mehlich-3 method (M3). The agri-environmental P saturation index $(P/A1)_{M3}$ was calculated. Phosphorus accumulation was observed in OG (0-5 cm) as a result of long-term manure applications. Repeated applications of organic fertilizers can impact the long-term buildup of soil P, thus decreasing soil P variability and spatial dependence in permanent grasslands. A soil sampling strategy focusing on the 0-5 cm layer should be retained in permanent grasslands for sustainable P recommendations in Eastern Canada.

2.4. Introduction

Phosphorus (P) is an essential nutrient for most crops, including associated plants of grassland communities. When it is applied in excess of crop requirements, P can accumulate at the soil surface, leading to P runoff via soil erosion and to P contamination of the environment, including eutrophication of surface waters such as lakes (Stroia et al., 2007; Jordan et al., 2000; Sun et al., 2015). To prevent this P contamination in the environment, the degree of P saturation of soils (%) was calculated from the ratio of phosphorus extracted with ammonium oxalate (Pox) on the maximum retention capacity of the P (CRP) (Maguire et al., 2002). The critical P saturation value of 25% has been defined as an international benchmark intended to prevent the risk of P loss and water contamination increases (Breeuwsma and Silva, 1992). Grasslands are defined as terrestrial ecosystems dominated by herbaceous and shrub vegetation (Obermeier et al., 2017, Zhou et al., 2017) and are distributed according to landscape and site characteristics (soil, climate, water availability) and farm structures. Permanent grasslands include grazed pastures and grasslands harvested for hay and silage, or a combination of the two, and are located primarily in Western Europe, including the United Kingdom and Ireland (Higgins et al., 2017). Canadian grassland fields covered an area of 1.96 million ha in 2016 (Statistics Canada, 2016). In the Eastern Canadian context, grassland cultivation represents an important economic opportunity for local farmers to increase their revenue. For instance, 50% of agricultural land was under intensive permanent grassland cultivation in the province of Quebec, producing 6.3 million tons of hay and generating a total economic profit of $19 M in 2016 (MAPAQ, 2018). Permanent grasslands and annual crops with reduced tillage represent crop systems with less risk of runoff and soil erosion, with respect to continuous annual crops under traditional systems with conventional tillage (CRAAQ, 2010). These two common crop systems were traditionally cultivated by grassland farmers, owing to their high economic revenue.

Moreover, P agricultural inputs were provided mainly from organic fertilizers (manure and slurry), mineral fertilizers, and other waste fertilizers in Eastern Canada (CRAAQ, 2010). Owing to their low economic costs, organic fertilizers represent 63% of total P sources, representing two-thirds of the total P applied by local farmers on agricultural lands (Hébert et al., 2007), including grassland fields. Over a long period, agricultural farms produced more than 31 million tonnes of organic manures, which were applied in agricultural fields (CRAAQ, 2010) and represented more than 95,000 tonnes of P2O5 (MDDELC, 2017). Thus, grassland fields were intensively fertilized using organic manures with high P concentrations (CRAAQ, 2010), increasing the environmental risk. Furthermore, in Eastern Canada, the soil sampling strategy for P fertilizer application is currently at 0-17.5 cm- depth in agricultural fields (CRAAQ, 2010). In 2012, total P loads in Eastern Canadian rivers were estimated at approximately 3800 tons of total P per year (Patoine et al., 2017). Thus, agri-environmental management of soil P in grassland fields should be based on scientific knowledge and understanding of soil-available P to develop better P fertilization strategies.

Soil P availability is defined as the total amount of soil P likely to end up in the soil solution as orthophosphate ions during a growing crop cycle (Frossard et al., 2004). In North America, soil-available P may be assessed for acidic soils using M3 extraction as an official method in Eastern Canada (Mehlich, 1984). From this method, two main P indices were developed, specifically (1) P-Mehlich-3 (PM3), an agronomic indicator for acidic mineral soils in Eastern Canada and USA, and (2) (P/Al)M3, an important agri-environmental P indicator, which estimates soil P saturation to allow accurate P fertilizer recommendations and which integrates agronomical aspects and environmental risks. In the province of Quebec, the critical environmental threshold for coarse soil corresponds to the value of 15% (CRAAQ, 2010). These critical saturation indices were incorporated into the government's environmental regulations related to agricultural lands (CRAAQ, 2010; MDDELC, 2017). In 2002, the national Agricultural Operations Regulation (REA) was adopted by the province of Quebec, establishing that the threshold level for P saturation (P/Al)3 corresponds to 13.1% in agricultural farms, including grassland fields (MDDELC, 2017). These critical saturation indices were incorporated into the province of Quebec's government environmental regulations related to agricultural lands (CRAAQ, 2010; MDDELC, 2017).

Furthermore, the concept of balanced fertilization underpins the sustainable use and management of soil P under intensive cropping systems. To achieve this goal, it is important to have a better understanding of the spatial variability of soil P in grassland fields, as this can help in improving the profitability and sustainability of agricultural businesses through the economic and rational use of P fertilizers, reducing P losses by runoff, and protecting the environment. Understanding spatial variability of soil P in grassland fields may be mainly based on two strategies: (1) variable rate applications (VRA) and (2) the use of management zones (MZs) in the context of precision agriculture. The VRA approach manages the variability in a factor based on a prescription map obtained with georeferenced data and geostatistical analysis, while the MZs approach delineates smaller homogeneous units in which uniform management can be applied using variable management among the different units (Cambouris et al., 2014). Those two approaches can help to reduce the losses of P in the environment by applying the P at the right place.

Many studies have been carried out on the spatial variability of soil P on a regional or territorial scale (Jia et al., 2011; Roger et al., 2014; Sun et al., 2015; Vasu et al., 2017) and at the watershed scale (Nolan et al., 2007; Yuan et al., 2009; Wilson et al., 2016). Most of these studies focused on planning and mapping soil- available P with a view to preventing and assessing the risk of P-related environmental pollution. However, there are no studies conducted investigating spatial variability of soil-available P in grassland fields under intensive production in Eastern Canada. Thus, this study is, to our knowledge, the first one on the spatial variability of soil P in grassland fields in Eastern Canada. The main goal of this study was to evaluate the spatial variability of soil-available P in grassland fields managed under intensive farming systems. More specifically, the objectives were to: (1) investigate field-scale variability of soil P indices and other selected soil chemical properties (soil pHwater, total carbon, AlM3, CaM3, FeM3), and (2) study the relationships between soil P indices (referred to as PM3 and (P/Al)M3) and selected soil chemical properties for two contrasting grasslands.

2.5. Materials and Methods

2.5.1. Study Site and Soil sampling

The study was conducted in the Bras d'Henri watershed (Chaudière-Appalaches region) located 30 km from Quebec City, QC, Canada. Two commercial fields were selected to represent different grassland management options: a young grassland field (46° 48'N; 71° 22'W, 2.35 ha), referred to as the YG field, and an old grassland field (46° 48'N; 71° 21'W, 2.47 ha), referred to as the OG field (Figure 2-1). The YG field was managed for corn (*Zea mays* L.; 2006; 2010) and soybean (*Glycine max*; 2007; 2011) rotation with 2 years of grassland (i.e., 2004-2005, 2008-2009 and 2012-2013) for 10 years. The YG field was planted in early May 2012. The OG field was used as a permanent pasture for 10 years (2004-2013). The YG field was fertilized with cattle manure and was harvested three times per year (mid-June, mid-August and late September). The OG field was fertilized with organic manure (e.g., calf slurry, pig manure and cattle excreta) and was harvested twice per year (mid-June and mid-August). Crop management and grassland fertilization practices were based on local recommendations for the province of Quebec (CRAAQ, 2010).

Both soils were classified as humo-ferric podzols (Lamontagne et al., 2010) and attributed to the Beaurivage soil series (Lamontagne et al., 2010). Soil surface texture for the YG and OG fields varied from sandy loam to loamy sand. They were classified to the same G3 soil texture group (i.e., texture with clay ≤ 20%) utilized in the province of Quebec recommendations (CRAAQ, 2010). Both fields were moderately well-drained. The annual winter mean temperature is -9.4° C, and the mean growing season temperature is 16.6° C (Environment Canada, 2016). Mean annual precipitation is 1120 mm, of which 320 mm is snow (Lamontagne et al., 2010).

The main grass species consisted of *Ladino* white clover (*Trifolium repens*) cultivated in association with timothy (*Phleum pratense* L.) in the YG field and tall fescue (*Festuca arundinacea* Schreb.) in the oG field. In 2004-2005, 2008-2009 and 2012-2013, cattle manure was applied and incorporated twice a year in the first 5 cm using a low-ramps spreading system in the YG field at the rate of 18,700 L ha^{-1} , equivalent approximately to 20 kg P ha^{-1} year^{-1} . Cattle manure was applied in spring and later after the first cut (mid-June).

In the OG field, calf and pig manures were applied at the rate of 30 Mg ha$^{\wedge 1}$ and 20 Mg ha$^{\wedge 1}$, respectively, giving approximately 67 kg P ha "1 year$^{\wedge}1$ over 10 years based on reference values of P published in CRAAQ (2003). At the beginning of May and after the first cut (mid-June), organic manures were surface applied in the OG field using a low-ramps spreading system.

An intensive soil sampling using a grid of 16 m by 16 m, including three cross-shaped designs with several inter-points at about 4 m apart, was performed in spring 2013, providing 151 and 149 georeferenced sampling points for the YG and OG fields, respectively (Figure 2-1). A composite soil sample of four cores was taken within a radius of 1 m around each sampling point at the two soil depths (0-5 cm and 5-20 cm) using a 0.05 m-diameter Dutch auger (Eijkelkamp company, Giesbeek, Netherlands). Thus, a total of 302 and 298 soil samples were collected from the YG and OG fields, respectively.

Soil samples were air-dried, ground and sieved through a 2 mm sieve for characterization. Soil pH (1:1 water) was measured, according to Hendershot et al. (2008). Soils were extracted with a soil solution ratio of 1:10 using Mehlich-3 solution (Ziadi and Tran., 2008), and the concentrations (mg kg^{-1}) of P_{M3} , iron (Fe_{M3}), calcium (Ca_{M3}) and aluminium (Al_{M3}) in the extract were determined by inductively coupled plasma optical emission spectroscopy (ICP-OES; Model, 4300DV, PerkinElmer, Shelton, CT, USA). The soil P agri- environmental indicator (P/Al)$_{Mi3was}$ then calculated. Total carbon (TC, g kg^{-1}) content was measured using an Elementar Vario MAX CN analyzer (Elementar Analysensysteme GmbH, Hanau, Germany).

2.5.2. Statistical and Geostatistical Analyses

Descriptive statistics were carried out using SAS Software version 9.4 (SAS Institute, 2010) to investigate soil pH_{water}, P_{M3}, Al_{M3}, (P/Al)$_{M3}$, Fe_{M3}, Ca_{M3} and TC and their spatial variability. The minimum, maximum, mean, coefficient of variation (CV) and standard deviation of the mean (SDM) were determined for each of the soil layers from each field. The CV of the different soil chemical properties was classified based on the approach of Nolin and Caillier (1992) using five classes: (1) low (CV < 15%); (2) moderate (15% < CV < 35%); (3) high (35% < CV < 50%); (4) very high (50% < CV < 100%); and (5) extremely high (CV > 100%).

Pearson's correlation coefficients (r) between soil P indices (P_{M3}, (P/A1)$_{M3}$) and other soil chemical properties (Al_{M3}, Ca_{M3}, Fe_{M3}, TC, and pH_{water}) were also calculated for each soil layer using SAS. Data were analysed using the CORR and univariate procedures. Multiple regression equations were generated for the soil P indices for each soil depth.

Geostatistical analyses were performed for all soil parameters. Data were achieved for each of the soil layers using the GS+ version 9 software (Webster and Oliver, 2007) geostatistical computations and model validations. The spatial structure of the different soil chemical properties was evaluated via isotropic and anisotropic semivariograms. Experimental semivariograms, the main component of kriging, are an effective tool for assessing spatial variability (Wu et al., 2009). Semivariogram parameters for each theoretical model (spherical, exponential and Gaussian) were generated. The corresponding nugget (c_0), partial sill (C), sill (C0 + C), and range values of the best-fitting theoretical model were calculated. The partial sill ratio (C/(C0 + C)), expressed as the percent of total semivariance, was used to define the spatial dependency or structure of the soil variables for the two fields. Semivariograms with a partial sill ratio of < 25%, 25% to 40%, 40% to 60%, 60% to 75%, or > 75% were considered to have a low, low-moderate, moderate, strong-moderate, or strong spatial dependency, respectively (Whelan and Mc Bratney, 2000). After the suitable theoretical model for each data set and the corresponding semivariogram parameters were selected, spatial variability maps were generated using 1 m × 1 m block kriging interpolation. Kriging map reliability was evaluated using cross-validation analysis (R^2 cv) (Kravchenko et al., 2002).

2.6. Results

2.6.1. Descriptive Statistics of Soil Phosphorus Indices and Other Chemical Properties

For the YG field, the mean values of soil P indices and other soil chemical properties were similar in both soil layers (0-5 cm and 5-20 cm), with the exception of Ca_{M3} (Table 2-1). Overall, the CVs ranged from 5% to 81%, with the highest obtained for the $(P/Al)_{M3}$, and were classified as follows: low for soil pH_{water} and Al_{M3}, moderate for Fe_{M3}, high for Ca_{M3} and TC, and very high for the two soil P indices.

For the OG field, soil P indices and other soil chemical properties were higher in the 0-5 cm layer than in the 5-20 cm layer, with the exception of Al_{M3}, for which the mean concentrations were similar (Table 2-1). On average, the CVs ranged from 6% to 64% in both soil layers, with the highest value obtained for the $(P/Al)_{M3}$, and were classified as follows: low for soil pH_{water} and Al_{M3}, moderate for Fe_{M3} and TC, and high for Ca_{M3}. The CVs of the soil P indices varied from high to very high in both soil layers. The intensity of variability of the soil P indices was higher in the YG compared to OG for both soil layers.

2.6.2. Geostatistical Parameters of Soil Phosphorus Indices and Other Chemical Properties

In the YG field, soil P indices and the other soil chemical properties (except for Ca_{M3}) were fitted with geostatistical spherical models, and pure nugget models (PNs) were used for Ca_{M3} in both soil layers. Spatial ranges varied from 28 to 81 m. Spatial ranges of soil chemical properties extended beyond the 16 m × 16 m sampling grid, indicating that the sampling grid used is appropriate. Spatial dependence ratios of soil P indices and other soil chemical properties ranged from 58% to 100% for the two soil layers and were classified according to Whelan and McBratney (2000) as follows: Fe_{M3} ranged from moderate to strong-moderate, pH_{water} and soil P indices were classified as strong-moderate, and TC and Al_{M3} were classified as strong (Table 2-2).

In the OG field, soil P indices and the other soil chemical properties (except for Al_{M3} and Ca_{M3}) were fitted with geostatistical spherical models, and PNs were used for Ca_{M3} and for Al_{M3} in the 0-5 cm soil layer (Table 2-2). Spatial ranges varied from 27 to 76 m and extended beyond the 16 m × 16 m sampling grid, indicating that the sampling grid used is appropriate. Spatial dependence ratios of soil P indices and the other soil chemical properties ranged from 26% to 69% in both soil layers. According to Whelan and McBratney (2000), the spatial dependence of Al_{M3} was low-moderate in the 5-20 cm layer. Spatial dependence values were moderate for soil pH_{water} and Ca_{M3}, strong-moderate to moderate for Fe_{M3}, low-moderate to moderate for soil P indices, and strong-moderate for soil TC in both soil layers (Table 2-2). Spatial dependence of soil P was lower under OG compared to YG for both soil layers.

In the YG field, cross-validation coefficients (R^2_{cv}) ranged from 0.31 to 0.54 for soil chemical properties and from 0.31 to 0.40 for soil P indices. In the OG field, R^2_{cv} were generally lower. The R^2_{cv} values ranged from 0.20 to 0.36 for the soil P indices. High R^2_{cv} values ($R^2_{cv} > 0.6$, Perron et al., 2018) indicate a good fit for kriged map reliability. Consequently, the R^2_{cv} values obtained for the YG field may indicate relative mean fits for map reliabilities for most soil chemical properties, including soil P indices, while the fits were weaker for map reliabilities of soil chemical properties in the OG field.

2.6.3. Spatial Distribution of Soil Phosphorus Indices and Other Chemical Properties

The P_{M3} and $(P/Al)_{M3}$ spatial distribution maps showed visual similarities for each soil layer studied (Figure 2-2 a,b,g,h and Figure 2-3 a,b,g,h). In the YG field, spatial distribution values of soil P indices, TC and Fe_{M3} were relatively similar in both soil layers (Figure 2-2a,b,d,g,i,j). Spatial distribution values of Al_{M3} were lower in the 0-5 cm soil layer compared to the 5-20 cm soil layer (Figure 2-2c,i).

However, in the OG field, spatial distribution values of soil P_{M3}, $(P/Al)_{M3}$, TC, and Fe_{M3} were higher in the 0-5 cm soil layer than in the 5-20 cm soil layer (Figure 2-3a,b,d,e,g,h,j,k). For instance, spatial distribution values of these soil chemical properties varied from 46 to 232 mg kg^{-1}, 2.8% to 13%, from 20 to 55 g kg^{-1} and from 135 to 324 mg kg^{-1} against 15-139 mg kg^{-1}, 1.1% to 8%, 20-40 g kg^{-1}, 135-297 mg kg^{-1} in 0-5 cm and 5-20 cm OG soil layers, respectively. In the OG field, no visual similarities were noted between Al_{M3} 0-5 cm and 5-20 cm or between Ca_{M3} 0-5 cm and 5-20 cm. This was not surprising, as the PN models were obtained with the semivariogram analysis.

In general, spatial distribution maps also revealed visual associations between soil P indices and other soil elements, indicating the influence of these different soil chemical properties on the soil-available P. For instance, in the YG field, the AI_{M3} and TC concentrations from the soil layers were inversely related to the soil P indices (Figure 2-2a-d,g-j).

In the OG field, the $A1_{M3}$ concentrations in the OG soil layer (5-20 cm) were inversely related to the soil P indices in OG (5-20 cm) (Figure 2-3g-i). The high Ca_{M3} concentrations correspond to soil P accumulation regions in the OG soil layer (5-20 cm), confirming P_{M3}-Ca_{M3} accumulation (Figure 2-3g,h,l). These results showed that AI_{M3} and TC content might influence soil P retention capacity in both YG soil layers, whereas AI_{M3}, Fe_{M3} and Ca_{M3} may affect soil P in both OG soil layers.

2.6.4. Relationships Between Soil Phosphorus Indices and Other Chemical Properties

In the YG field, Pearson's correlation coefficients (r) with the highest significance were observed between P_{M3} and AI_{M3}, followed by TC in both soil layers (Table 2-3). For $(P/AI)_{M3}$, the highest r values were obtained with TC in both soil layers. In the OG field, significant relationships were observed between P_{M3} and AI_{M3}, Ca_{M3} and soil pH_{water} in both soil layers. However, the highest r was observed between $(P/AI)_{M3}$ and Ca_{M3} in each soil layer (Table 2-3). Thus, soil P_{M3} in the YG field may be influenced significantly by Al oxides and hydroxides (soil Al), soil organic matter (TC) and soil solution (soil pH_{water}), whereas in the OG field, Al oxides/hydroxides (soil Al), calcium carbonate (soil Ca), soil solution (soil pH_{water}) and iron oxides/hydroxides (soil Fe) may have a significant influence on soil P accumulation.

Multiple regression equations were generated for the soil P indices in the YG and OG soil layers to investigate the influence of these soil chemical properties (Table 2-4). In the YG field, the regression coefficients (R^2) were very significant, ranging from 0.789 to 0.893, which means that 79% to 89% of soil-available P variations may be explained by AI_{M3}, TC and soil pH_{water} in both soil layers. Similarly, in the OG field, the R^2 varied between 0.532 and 0.659, indicating that 53% to 66% of soil-available P accumulations in both soil layers may be affected significantly by AI_{M3}, soil pH_{water}, Ca_{M3} and Fe_{M3}. Consequently, permanent grassland may accumulate soil P while significantly affecting the effect of these soil chemical properties.

2.7. Discussion

2.7.1. Soil Phosphorus Stratifications and Its Environmental implications

In this study, soil P accumulation was observed in the 0-5 cm OG soil layer, mainly due to multiple applications of organic fertilizers (calf slurries, pig manure, and cattle excreta) over a long time period. Indeed, soil P accumulation in the 0-5 cm OG top layer may also be due to the elimination of soil disturbance (no tillage, compared to the conventional tillage) under permanent grasslands.

Many studies (Fu et al., 2000; Darilek et al., 2010; Roger et al., 2014) have shown that soil P accumulation in the surface layer from intensively cultivated lands is linked to increased applications of P fertilizers, organic inputs, soil land-use conversion and management. For instance, Sun et al. (2015) reported that significant increases in soil-available P (P Olsen) were attributed to intensive local applications of swine manure over a 15-year period (1982-1997) in the Rugao region (China). Other studies (Bennett et al., 2005; Roger et al., 2014) found that soil P indices (P-Bray I, P-ammonium-acetate-EDTA (P_{AAE}): referred to as the Swiss reference method) varied significantly in different land uses and crop systems, including permanent grassland fields.

This study revealed that repeated application of organic fertilizers raised soil P accumulation in the first five soil centimeters, increasing the potential agri-environmental risk for P pollution in permanent grasslands. The mean $(P/A1)_{M3}$ indicator raised up to twice (7%) in the 0-5 cm OG soil layer, relative to the 0-5 cm YG soil layer (3%). These results have several environmental implications for the sustainable management of soil P in grassland fields. Indeed, the $(P/AI)_{M3}$ mean values obtained from the OG site were close to the national critical threshold P saturation (15%) elaborated by the Province of Quebec for coarse-textured soils, meaning that $(P/AI)_{M3}$ values higher than 15% increased up to water pollution or groundwater contamination in this Eastern Canadian region.

Moreover, these results also revealed potential agri-environmental impacts for the management of permanent grasslands in the Eastern Canadian region. According to the national Agricultural Operations Regulation (REA) adopted by the province of Quebec in 2002, these obtained results were close to the threshold level (13%) for P saturation (P/Al)$_{M3}$ for agricultural lands, including grassland fields. Soil P stratification or buildup is an important agri-environmental concern because the excess surface P may be lost in runoff (Sharpley and Smith, 1994). At the same time, however, lower concentrations at depth in the rooting zone may reduce crop yields (Lupwayi et al., 2006).

2.7.2. Variability of Soil Phosphorus Indices and Other Chemical Properties

The highest CVs were obtained for soil P indices and the lowest CVs for soil pH measurements. These results were similar to other studies (Arrouays et al., 1997; Cambouris et al., 2006; Di Virgilio et al., 2007; Belanger et al., 2016;Vasu et al., 2017). A low CV for soil $_{pHwater}$ may be due to the logarithmic scale of pH measurement (Farooque et al., 2012) or may be attributable to management practices, including regular soil liming activities in intensive agriculture (Nolin et al., 1991). West et al. (1989) reported that a high CV (47%) for soil-available P could be explained by chemical forms of measured P that are highly labile.

Many studies (McCormick et al., 2009; Fu et al., 2013; Fu et al., 2013; Vasu et al., 2017) have likewise reported very high CVs for soil-available P (CV > 50%), particularly on Irish soils under intensive grassland farming. McCormick et al. (2009) reported a CV of 57% (n = 334) for a 50-ha parcel of land encompassing 15 small grassland fields ranging in size from 0.94 ha to 5.65 ha in Northern Ireland. Fu et al. (2013) obtained a CV of 68% (n = 537) for a 52-ha grassland field located in southeastern Ireland.

This study shows that a reduction in the CV of soil P was observed in permanent grasslands. For example, Jia et al. (2011) reported a large decrease in the CV of soil P from 151% to 55% after 25 years of intensive farming practices in China. However, some authors (Nolin et Caillier., 1992; Jia et al., 2011; Fu et al., 2013; Roger et al., 2014) found that long-term fertilization practices or intensive land-use changes, including permanent grasslands, may influence CV values of soil P indices. Roger et al. (2014) reported extremely high CV values for soil-available P (PAAE) at permanent grasslands sites (CV = 107%) compared to mountain pasture sites (CV = 87%) and cropland sites (CV = 70%). From their study, the CVs of soil P saturation [(Pox/(Al+ Fe)ox) × 100] ranged from 35% to 38%. In general, CV values are good indicators of the intensity of soil variability, but not of the nature of this variability (structured or randomized variability), nor do they provide an accurate model of soil variability (Cambouris et al., 2006).

2.7.3. Spatial Dependence of Soil Phosphorus Indices and Other Chemical Properties

Spatial dependence (R = (C/C0 + C)) describes the spatial structure of a soil property (Quenum et al., 2012). Some authors (Cambardella et al., 1994; Goovaerts, 1998; Vasu et al, 2017) have indicated that soil chemical properties with strong spatial dependence (R > 75%) are mainly influenced by soil intrinsic factors (soil formation factors, parental material, soil types or series), those with low spatial dependence (R < 25%) are strongly influenced by extrinsic factors (agronomic/cultural or fertilization practices, intensive use of organic fertilizers or mineral amendments), and those with moderate spatial dependence (25% < R < 75%) result from the interaction between intrinsic and extrinsic soil factors, which codetermine the soil property analysed. The PNs models are characterised by a spatial structure smaller than the sampling grid because of an inappropriate sampling scale (Arrouays et al., 1997).

The spatial dependence of soil P indices in both YG and OG soil layers was moderate (31% to 71%) as a result of the interaction between soil intrinsic and extrinsic factors, such as intensive fertilization or farming practices. Many authors (Chien et al., 1997; Chang et al., 1999; Shi et al., 2000; Whelan and Mc Bratney., 2000; Guo-Shun et al., 2008; Jia et al., 2011; Vasu et al., 2017; Cambouris et al., 2017) have found moderate spatial dependence for soil P, while others (Cambouris et al., 2006; Mc Cormick et al., 2009; Roger et al., 2014; Sun et al., 2015) have reported strong spatial dependence for soil P (R > 75%) in their respective studies conducted on different scales. The strong spatial dependence of soil-available P observed in these studies may be attributable to the strong influence of intrinsic factors, such as different soil types or soil series, on the spatial distribution of soil-available P at different agricultural, local, or regional scales.

Spatial dependence values for soil $_{pHwater}$ and $_{FeM3}$ in the YG and OG soil layers were moderate (41% to 74%), owing to the interaction between soil intrinsic and extrinsic factors. Arrouays et al (1997), Cambouris et al (2006), and Di Virgilio et al (2007) obtained similar results for soil $_{pHwater}$. Spatial dependence values for soil $_{AlM3}$ and TC ranged from strong to moderate in the YG and OG soil layers, respectively. Spatial dependence values obtained for $AIM3$ in YG soil layers were lower than those reported by Cambouris et al. (2006). For soil TC, Kokulan et al. (2018) and Arrouays et al. (1997) reported similar results for YG and OG soil layers (0-5 cm and 5-20 cm). Consequently, $A1M3$ and TC in both YG soil layers may result from the strong influence of soil intrinsic factors, whereas $AIM3$ and TC in both OG soil layers result from the interaction between soil intrinsic and extrinsic factors.

This study also revealed that a combination of intensive fertilization practices over a longer period reduced the spatial dependence of soil P in both OG soil layers. This may be explained by the increasing influence of soil extrinsic factors, such as P-rich fertilization, which modified the long-term spatial structure of soil P in the OG field. Other authors (Jia et al., 2011; Fu et al., 2013; Sun et al., 2015) reported a reduction of spatial dependence for soil P on Chinese and Irish soils at different regional and agricultural scales. These studies demonstrated that a weak spatial dependence of P distribution could be explained by various intensive agricultural farming practices (i.e., increasing fertilization, long-term P-rich mixed fertilization, and strong intensification of sheep grazing pasture activities).

From an agronomic perspective, the reduced spatial structure of soil P may affect the use of delineated MZs for sustainable management of soil P in grassland fields. A strong spatial structure can have a positive effect in terms of delineating MZs, while a weak spatial structure makes it difficult to use MZs, owing to the lack of spatial dependence.

The use of MZs is an agricultural approach based on the existence of within-field spatially structured soil and crop variability (Cambouris et al., 2006; Farooque et al., 2012). Subdividing a field into MZs is an effective way of understanding or controlling spatial variability of soil and crop properties in order to optimize soil nutrients and crop management (Perron et al., 2018). Delineation of MZs has been tested successfully for various soil chemical properties (Zebarth et al., 2009; Cambouris et al., 2014) and crop fields (Cambouris et al., 2006; Farooque et al., 2012), including old grassland fields (Belanger et al., 2016), as well as different field sizes (Farooque et al., 2012; Perron et al., 2018). However, there are few examples of the use of this approach with the aim of optimizing the management of P variability in grassland fields.

The present study on soil P variability was carried out in 2.5-ha fields. Consequently, it would be mitigated to delineate MZs for these fields due to (1) the reduced spatial structure of soil P, which is caused by an increase in extrinsic soil factors, and (2) the size of the fields with respect to the crop studied. Further research on larger fields is needed to better define P spatial structure in grasslands.

2.7.4. Spatial Distribution of Soil Phosphorus Indices and Other Chemical Properties

The R^2 cv values provide an indication of the reliability of spatial distribution maps. Weak R^2 cv values were obtained for soil P indices and other chemical properties in the OG field compared to the YG field. Use of R^2 cv values in grassland fields aims to evaluate the reliability of kriged spatial distribution maps of soil P indices for developing better agri-environmental strategies for sustainable P management in the context of precision agriculture. Thus, R^2 cv should also be considered as an important reliability factor for mapping soil P indicators in grassland fields to prevent P pollution. Some authors (Cambouris et al., 2017) reported lower R^2 cv values (0.0-0.20), while others (Vasu et al., 2017; Perron et al., 2018) found higher R^2 cv values (0.59- 0.62) for soil-available P. In this study, the lowest R^2 cv values obtained may be explained by the weak spatial structure of the soil P obtained, as suggested by Kravchenko (2003).

Despite their low reliability (R^2 cv < 0.4), spatial distribution maps of soil P indices (Figure 2-3a,b) confirmed that the soil P accumulation observed in the OG soil layer (0-5 cm) is attributable to the repeated application of organic fertilizers. Furthermore, in this study, high P concentration areas were found in the YG (81-129 mg kg^{-1} ; 5% to 8%) and OG (139-232 mg kg^{-1} ,8% to 13%) fields, probably due to uniform applications of P-rich organic fertilizers. The YG field was fertilized with cattle manure under corn and soybean rotation and included 2 years of grassland over a period of 10 years, whereas the OG field was fertilized for 10 years with multiple applications of organic fertilizers (e.g., calf slurries, pig manure, and cattle excreta).

Spatial distribution maps for soil $AIM3$ in both sites also confirmed the strong influence of high available Al content on soil P adsorption capacity in grassland soils in Eastern Canada. In a previous study, Tran et al (1990) highlighted this possibility in relation to Quebec agricultural soils. Taking into account soil $AIM3$ concentrations, these authors classified Quebec soils into three groups according to their capacity for soil P adsorption: (1) low capacity: $[A1]_{M3}$ < 1100 mg kg^{-1} ; (2) average capacity: 1100 mg kg^{-1} < $[A1]_{M3}$ < 1600 mg kg^{-1} ; and (3) high capacity $[Al]_{M3}$ > 1600 mg kg^{-1} . Furthermore, this study revealed that grassland fields in Eastern Canada have an average-high capacity for soil P adsorption. From these results, the spatial distribution of soil Al_{M3} could be evaluated as a potential indicator for soil P adsorption in grassland fields. This approach would aim to implement sustainable management of P as mineral or organic fertilizer, based on VRA of P fertilizer for preventing pollution. For instance, other studies (Quenum et al., 2012) found that $A1_{M3}$ could be considered as an important indicator of maximum P adsorption capacity for a VRA of P fertilizer in the agricultural fields from Eastern Canada.

The spatial distribution maps for soil P indices and other soil elements ($A1_{M3}$, TC, Fe_{M3} and Ca_{M3}) in YG and OG soil layers developed in this study confirm the observed correlation, particularly between soil P indices and soil Ca_{M3} and Fe_{M3} in the 5-20 cm OG soil layer. The southern area of the OG field (5-20 cm) may reflect soil P accumulation linked to Fe oxides or hydroxides and neoformation of Ca phosphates as a result of longterm fertilization (Figure 2-3k,l).

2.7.5. Relationships Between Soil Phosphorus Indices and Other Chemical Properties

The results showed differences between YG and OG. Soil P may accumulate or increase in OG, and there may be significant changes of different soil parameters (Al_{M3}, TC, Ca_{M3}, Fe_{M3}) on soil P indices, which are probably due to a decrease in soil pH. Decreasing soil pH is common in intensively farmed soils due to heavy applications of inorganic N fertilizers (Wang et al., 2008; Darilek et al., 2009; Darilek et al., 2010). It is well known that soil pH plays an important role in P solubility and adsorption processes, such as P adsorption onto soil oxides, minerals and soil organic matter (Havelin et al., 2005; Devau et al., 2010). For instance, Fu et al. (2013) mentioned a significant relationship between soil-available P and soil pH (r = 0.69, n = 396, p < 0.01) in a grassland field used for grazing for 16 years. In this study, we found significant relationships between soil P indices and soil pH (r = 0.43 for P_{M3}, p < 0.001; r = 0.46 for $(P/A1)_{M3}$, p < 0.001) in the 0-5 cm OG soil layer.

Results also show that soil P accumulation in the OG soil layers (0-5 cm and 5-20 cm) may be related to soil-available concentrations of Al, Ca and Fe resulting from a decrease of soil pH. Many studies (Sharpley and Smith., 1985; Guo et al., 2000; Motavalli and Milles, 2002; Darilek et al., 2010; Sun et al., 2015) reported that various agricultural and farming practices affect soil P accumulation. Other studies (Kuo et al., 2005; Devau et al., 2011) observed relationships between soil P buildup and Fe/Al/-P, attributed to the intensive use of organic and inorganic P fertilizers in cultivated acid soils.

Positive relationships observed between soil P_{M3} buildup and soil Ca_{M3} in the OG field may be indicative of soil P accumulation as Ca-bound P or neoformation of Ca phosphates after long-term P fertilization. This possibility has previously been advanced by several authors (Darilek et al., 2010; Eriksson et al., 2016). These earlier studies showed that the P-Ca relationship also depends on the relationship between soil pH and Ca. In the present study, we reported a significant relationship between soil pH and soil Ca_{M3} in the 0-5 cm (r = 0.65; p < 0.001) and 5-20 cm (r = 0.57; p < 0.001) OG soil layers.

2.7.6. Agronomic and Environmental Implications

Repeated applications of organic fertilizers and other agronomical practices such as no-till soil may increase soil P accumulation in the first 5 cm of soil in agricultural fields (Cade-Menum et al., 2010; Abdi et al., 2014; Cambouris et al., 2017), including grasslands (Belanger et al., 2016). This points to the importance of developing an effective soil sampling optimization strategy to provide better guidance for practical grassland management. Fertility management in grasslands is related to sampling depth. Although some studies (Haygarth et al, 1998; Watson et al., 2007;Watson and Matthews, 2008) demonstrated that the 0-5 cm layer might provide the majority of N and P in grassland fields, a deeper soil profile may be important to sustaining healthy grasslands.

Soil sampling strategy in grassland fields represents a balancing act between fertility management and minimizing environmental impacts (i.e., P pollution, water quality). The current 0-17.5 cm sampling strategy may be more suitable for yield optimization, while the 0-5 cm sampling strategy would be better for more sustainable P recommendations in grassland fields to reduce P pollution in the environment. According to the results of this study, we recommend the use of a soil sampling strategy focusing on the 0-5 cm layer in permanent grassland fields, given the high risk of soil P stratification associated with P-rich fertilization, instead of the soil sampling strategy (0-17.5 cm) currently used in Eastern Canada, in order to derive more sustainable P recommendations for grasslands. Furthermore, other studies (Stroia et al., 2007; Watson and Matthews, 2008;McCormick et al., 2009; Stroia et al., 2011) recommended soil sampling strategies in grassland fields, focusing on the 0-5 cm, 07.5 cm or 0-10 cm layers in other countries.

In this study, reduction of soil P variability under permanent grasslands was observed, inversely to other previous studies achieved within grassland fields (West et al., 1989; Higgins et al., 2017). Therefore, sustainable management of soil P in grassland fields should aim to better control within field soil P variability. Spatial distribution maps of soil P indices should be used to guide variable-rate P fertilizer applications in order to prevent under- and over-fertilization with P in grassland fields.

Furthermore, new technology advances (reflectance and evaluation of P) and the democratization of prices for VRA will make possible the potential use of this technique in the grasslands in the near future. Indeed, the implementation of local or environmental policies about agri-environmental and sustainable management of P will encourage grassland farmers to turn to these new technologies in the context of PA. Hence, although it is true that the additional economic costs can be high, the VRA approach would be an advantage or benefit for Eastern Canadian grassland farmers in reducing P losses in the environment.

Delineating MZs in grassland fields represents another option that can be used in the future to optimize the management of soil P variability in grassland fields. However, this agronomical practice has some limitations in adopting such an approach. These constraints are (1) a more reduced spatial structure of soil P, caused by a strong influence of extrinsic factors comparatively to the intrinsic factors, and (2) the field size with respect to the crop studied, which may influence the spatial distribution of MZs and their reliability. To overcome these constraints, MZs should be based on stronger soil P spatial structures obtained from reliable semivariograms, which are needed to calculate sampling intervals and develop an accurate site-specific application scheme for soil P in grasslands. In this context, field size is an important soil-related factor that needs to be taken into consideration. Further research on larger fields is needed to better define P spatial structure in grasslands.

2.8. Conclusions

This study represents, to our knowledge, the first one on the spatial variability of soil P in grassland fields in Eastern Canada. The main goal of this study was to evaluate the spatial variability of soil-available P in grassland fields managed under intensive farming systems, using descriptive statistics and geostatistical tools for accurate recommendations on soil sampling strategy, and developing potential future and sustainable approaches of P management in grassland fields.

In this study, repeated applications of organic fertilizers under grassland production with minimum soil disturbance increased soil-available P buildup [P_{M3}, $(P/A1)_{M3}$] at the 0-5 cm OG soil layer. This soil P accumulation increased the potential agri-environmental risk for P pollution in permanent grasslands. Under permanent grasslands, reducing variability and spatial dependence of soil P indices were observed in both soil layers (0-5 cm and 5-20 cm), as a result of the increasing influence of extrinsic soil factors, such as repeated applications of organic fertilizers over a long time. From an agronomic perspective, the reduced spatial structure of soil P may affect the use of delineated MZs for sustainable management of soil P in grassland fields. Despite their low reliability, spatial distribution maps of soil P indices should be used to guide variable-rate P fertilizer applications in order to prevent under- and over-fertilization with P in grassland fields.

In keeping with the results of this study, we recommend a soil sampling strategy focusing on the 0-5 cm layer in permanent grassland fields, given the high risk of soil P stratification caused by P-rich fertilization, as opposed to the soil sampling strategy (0-17.5 cm) currently used in Eastern Canada, in order to generate more sustainable P recommendations for grasslands. Moreover, this primary essay on the spatial variability of soil P was only restricted to two selected grassland fields. Further studies in other and larger grassland fields are required for a better understanding of the spatial variability of P to improve agri-environmental P fertilization for grassland fields.

2.9. Acknowledgements

Funding for this study was provided by Agriculture and Agri-Food Canada. We thank Claude Lévesque, Sylvie Michaud, and Marc Duchemin for their field assistance and technical support.

2.10. References

Abdi, D., Cade-Menun, B.J., Ziadi, N., and Parent, L.E. (2014). Long-term impact of tillage practices and phosphorus fertilization on soil phosphorus forms as determined by 31 P nuclear magnetic resonance spectroscopy. *Journal of Environmental Quality* 43, 1431-1441.

Arrouays, D.; Vion, I.; Jolivet, C.; Guyon, D.; Couturier, A.; Wilbert, J. (1997). Intraparcel variability of some sandy soil properties in the Landes de Gascogne (France): Consequences for agronomic sampling strategy. *Étude et Gestion des Sols* 4, 5-16.

Bélanger, G.; Cambouris, A.N.; Parent, G.; Mongrain, D.; Ziadi, N.; Perron, I. (2016). Biomass yield from an old grass field as affected by sources of nitrogen fertilization and management zones in northern areas. *Canadian Journal of Plant Science* 97, 53-64.

Bennett, E.M.; Carpenter, S.R.; Clayton, M.K. (2005). Soil phosphorus variability: Scale-dependence in an urbanizing agricultural landscape. *Landscape Ecology* 20, 389-400.

Breeuwsma, A.; Silva, S. (1992). Phosphorus Fertilisation and Environmental Effects in The Netherlands and the Po Region (Italy); *DLO The Winand Staring Centre: Wageningen,* The Netherlands.

Cade-Menun, B.J.; Carter, M.R.; James, D.C.; Liu, C.W. (2010). Phosphorus forms and chemistry in the soil profile under long-term conservation tillage: A phosphorus-31 nuclear magnetic resonance study. *Journal of Environmental Quality* 39, 1647-1656.

Cambardella, C.; Moorman, T.; Parkin, T.; Karlen, D.; Novak, J.; Turco, R.; Konopka, A. (1994). Field-scale variability of soil properties in central Iowa soils. *Soil Science Society of America Journal* 58, 15011511.

Cambouris, A.; Nolin, M.; Zebarth, B.; Laverdière, M. (2006). Soil management zones delineated by electrical conductivity to characterize spatial and temporal variations in potato yield and in soil properties. *American Journal of Potato Research* 83, 381-395.

Cambouris, A.N.; Zebarth, B.J.; Ziadi, N.; Perron, I. (2014). Precision agriculture in potato production. *Potato Research* 57, 249-262.

Cambouris, A.; Messiga, A.; Ziadi, N.; Perron, I.; Morel, C. (2017). Decimetric-Scale Two-Dimensional Distribution of Soil Phosphorus after 20 Years of Tillage Management and Maintenance Phosphorus Fertilization. *Soil Science Society of America Journal* 81, 1606-1614.

Chang, J.; Clay, D.E.; Carlson, C.G.; Malo, D.; Clay, S.A.; Lee, J.; Ellsbury, M. (1999) Precision farming protocols: Part 1. Grid distance and soil nutrient impact on the reproducibility of spatial variability measurements. *Precision Agriculture 1,* 277-289.

Chien, Y.-J.; Lee, D.-Y.; Guo, H.-Y.; Houng, K.-H. (1997). Geostatistical analysis of soil properties of mid-west Taiwan soils. *Soil Science 62,* 291-298.

CRAAQ. (2003). Guide de Référence en Fertilisation, 1st ed; *Centre de référence en agriculture et agroalimentaire du Québec: Québec, QC, Canada*; p. 294.

CRAAQ. (2010). Guide de Référence en Fertilisation, 2nd ed; *Centre de Référence en Agriculture et Agroalimentaire du Québec: Québec, QC, Canada*; p. 473.

Darilek, J.; Huang, B.; De-Cheng, L.; Zhi-Gang, W.; Yong-Cun, Z.; Wei-Xia, S.; Xue-Zheng, S.J.P. (2010). Effect of land use conversion from rice paddies to vegetable fields on soil phosphorus fractions. *Pedosphere* 20, 137-145.

Darilek, J.L.; Huang, B.; Wang, Z.; Qi, Y.; Zhao, Y.; Sun, W.; Gu, Z.; Shi, X.J.A. (2009). Changes in soil fertility parameters and the environmental effects in a rapidly developing region of China. *Agriculture, Ecosystems and Environment* 129, 286-292.

Devau, N.; Hinsinger, P.; Le Cadre, E.; Gérard, F. (2011). Root-induced processes controlling phosphate availability in soils with contrasted P-fertilized treatments. *Plant and Soil* 348, 203-218.

Devau, N.; Le Cadre, E.; Hinsinger, P.; Gérard, F. (2010). A mechanistic model for understanding root-induced chemical changes controlling phosphorus availability. *Annals of Botany* 105, 1183-1197.

Di Virgilio, N.; Monti, A.; Venturi, G. (2007). Spatial variability of switchgrass (Panicum virgatum L.) yield as related to soil parameters in a small field. *Field Crops Research* 101, 232-239.

Environment Canada (2013). *Canadian Climate Normals or Averages 1971-2000.* Available online:

http://www.climat.meteo.ec.gc.ca/climate normals/index e.html (accessed on 27 November 2016).

Eriksson, A.K.; Hesterberg, D.; Klysubun, W.; Gustafsson, J.P. (2016). Phosphorus dynamics in Swedish agricultural soils as influenced by fertilization and mineralogical properties: Insights gained from batch experiments and XANES spectroscopy. *Science of the Total Environment* 566, 1410-1419.

Farooque, A.A.; Zaman, Q.U.; Schumann, A.W.; Madani, A.; Percival, D.C. (2012). Delineating management zones for site specific fertilization in wild blueberry fields. *Applied Engineering in Agriculture 28, 57-70.*

Frossard, E.; Julien, P.; Neyroud, J.-A.; Sinaj, S. (2004). Le Phosphore Dans les sols: État de la Situation en Suisse: Le Phosphore Dans les Sols, les Fertilrais, les Cultures et L'environnement; *Office fédéral de l'environnement, des forêts et du paysage OFEFP:* Berne, Switzerland; p. 180.

Fu, B.; Chen, L.; Ma, K.; Zhou, H.; Wang, J. (2000). The relationships between land use and soil conditions in the hilly area of the loess plateau in northern Shaanxi, China. *Catena* 39, 69-78.

Fu, W.; Zhao, K.; Jiang, P.; Ye, Z.; Tunney, H.; Zhang, C. (2013). Field-scale variability of soil test phosphorus and other nutrients in grasslands under long-term agricultural managements. *Soil Research* 51, 503512.

Fu, W.; Zhao, K.; Tunney, H.; Zhang, C. (2013). Using GIS and geostatistics to optimize soil phosphorus and magnesium sampling in temperate grassland. *Soil Science* 178, 240-247.

Goovaerts, P. (1998). Geostatistical tools for characterizing the spatial variability of microbiological and physicochemical soil properties. *Biology and Fertility of Soils* 27, 315-334.

Guo, F.; Yost, R.; Hue, N.; Evensen, C.; Silva, J. (2000). Changes in phosphorus fractions in soils under intensive plant growth. *Soil Science Society of America Journal* 64, 1681-1689.

Guo-Shun, L.; Xin-Zhong, W.; Zheng-Yang, Z.; Chun-Hua, Z. (2008). Spatial variability of soil properties in a tobacco field of central China. *Soil Science* 173, 659-667.

Havlin, J.L.; Beaton, J.D.; Tisdale, S.L.; Nelson, W. (2005). Soil Fertility and Fertilizers: An Introduction to Nutrient Management; *Pearson Prentice Hall: Upper Saddle River,* NJ, USA.

Haygarth, P.; Hepworth, L.; Jarvis, S. (1998). Forms of phosphorus transfer in hydrological pathways from soil under grazed grassland. *European Journal of Soil Science* 49, 65-72.

Hébert, M.; Busset, G.; Groeneveld, E. (2008). Bilan 2007 de la Valorisation des Matières Résiduelles Fertilisantes; *Ministère du Développement durable, de ! Environnement et des Parcs: Québec, QC,* Canada; p. 14.

Hendershot, W.H.; Lalande, H.; Duquette, M. (2008). Soil reaction and exchangeable acidity. *In Soil Sampling and Methods of Analysis; ed. B.R. Taylor & Francis, Fl. Taylor & Francis, Boca Raton, FL,* 173-178.

Higgins, S.; Schellberg, J.; Bailey, J. (2017). A review of Precision Agriculture as an aid to Nutrient Management in Intensive Grassland Areas in North West Europe. *Advances in Animal Biosciences* 8, 782-786.

Jia, S.; Zhou, D.; Xu, D. (2011). The temporal and spatial variability of soil properties in an agricultural system as affected by farming practices in the past 25 years. *Journal of Food, Agriculture, and Environment* 9, 669-676.

Jordan, C.; McGuckin, S.; Smith, R. (2000) Increased predicted losses of phosphorus to surface waters from soils with high Olsen-P concentrations. *Soil Use and Management* 16, 27-35.

Kokulan, V.; Akinremi, O.; Moulin, A.P.; Kumaragamage, D. (2018). Importance of terrain attributes in relation to the spatial distribution of soil properties at the micro scale: A case study. *Canadian Journal of Soil Science* 98, 292-305.

Kravchenko, A. (2003). Influence of spatial structure on accuracy of interpolation methods. *Soil Science Society of America Journal* 67, 1564-1571.

Kravchenko, A.; Bollero, G.A.; Omonode, R.; Bullock, D. (2002). Quantitative mapping of soil drainage classes using topographical data and soil electrical conductivity. *Soil Science Society of America Journal* 66, 235-243.

Lamontagne, L.; Martin, A.; Nolin, M.C. (2010). Étude Pédologique du Bassin Versant du Bras d'Henri (Québec); Laboratoires de pédologie et d'agriculture de précision, Centre de recherche et de développement sur les sols et les grandes cultures, Service national d'information sur les terres et les eaux. *Research Branch, Agriculture and Agri-Food Canada: Québec, QC,* Canada; p. 188.

Kuo, S.; Huang, B.; Bembenek, R. (2005). Effects of long-term phosphorus fertilization and winter cover cropping on soil phosphorus transformations in less weathered soils. *Biology and Fertility of Soils* 41, 116-123.

Lupwayi, N.; Clayton, G.; O'Donovan, J.; Harker, K.; Turkington, T.; Soon, Y. (2006). Soil nutrient stratification and uptake by wheat after seven years of conventional and zero tillage in the Northern Grain belt of Canada. *Canadian Journal of Soil Science* 86, 767-778.

Maguire, R.O.; Sims, J.T. (2002). Measuring agronomic and environmental soil phosphorus saturation and predicting phosphorus leaching with Mehlich 3. *Soil Science Society of America Journal* 66, 20332039.

MAPAQ. (2018). Portrait-Diagnostic Sectoriel de L'industrie des Plantes Fourragères du Québec; *Ministère de l'Agriculture de l'Alimentation et des Pêcheries du Québec:* Québec, QC, Canada; p. 27.

McCormick, S.; Jordan, C.; Bailey, J. (2009). Within and between-field spatial variation in soil phosphorus in permanent grassland. *Precision Agriculture* 10, 262-276.

MDDELCC. (2017). Guide de Référence du Règlement sur les Exploitations Agricoles; *Ministère du Développement durable, de l'Environnement et de la Lutte contre les changements climatiques: Québec,* QC, Canada; p. 185.

Mehlich, A. (1984). Mehlich 3 soil test extractant: A modification of Mehlich 2 extractant. *Communications in Soil Science and Plant Analysis* 15, 1409-1416.

Motavalli, P.; Miles, R. (2002). Soil phosphorus fractions after 111 years of animal manure and fertilizer applications. *Biology and Fertility of Soils* 36, 35-42.

Nolan, S.; Little, J.; Casson, J.; Hecker, F.; Olson, B. (2007). Field-scale variation of soil phosphorus within small Alberta watersheds. *Journal of Soil and Water Conservation* 62, 414-422.

Nolin, M.; Caillier, M. (1992). Soil variability. II-Quantification et amplitude. *Agrosol* 5, 21-32.

Nolin, M.; Caillier, M.; Wang, C. (1991). Soil variability and sampling strategy in detailed soil surveys of the Montreal Plain. *Canadian Journal of Soil Science* 71, 439-451.

Obermeier, W.; Lehnert, L.; Kammann, C.; Müller, C.; Grünhage, L.; Luterbacher, J.; Erbs, M.; Moser, G.; Seibert, R.; Yuan, N. (2017). Reduced CO_2 fertilization effect in temperate C3 grasslands under more extreme weather conditions. *Nature Climate Change* 7, 137-141.

Patoine, M.; Hébert, S.; Simoneau, M.; d'Auteuil-Potvin, F. (2017). Phosphorus, Nitrogen and Suspended Matter Loads at the Mouths of Québec Rivers, 2009 to 2012; *Ministère du Développement durable, de l'Environnement et de la Lutte contre les changements climatiques, Direction générale du suivi de l'état de l'environnement: Québec, QC, Canada*; p. 25.

Perron, I.; Cambouris, A.N.; Chokmani, K.; Vargas Gutierrez, M.F.; Zebarth, B.J.; Moreau, G.; Biswas, A.; Adamchuk, V. (2018). Delineating soil management zones using a proximal soil sensing system in two commercial potato fields in New Brunswick, Canada. *Canadian Journal of Soil Science* 98, 724-737.

Quenum, M.; Nolin, M.; Bernier, M. (2012). Digital mapping of the maximum phosphorus sorption capacity of soils at the agricultural plot scale using auxiliary variables. *Canadian Journal of Soil Science* 92, 733-750.

Roger, A.; Libohova, Z.; Rossier, N.; Joost, S.; Maltas, A.; Frossard, E.; Sinaj, S. (2014). Spatial variability of soil phosphorus in the Fribourg canton, Switzerland. *Geoderma* 217-218, 26-36.

SAS Institute. (2010). *SAS User's Guide. Statistics. Version 9.3; SAS Institute: Cary,* NC, USA.

Sharpley, A.; Smith, S. (1985). Fractionation of Inorganic and Organic Phosphorus in Virgin and Cultivated Soils. *Soil Science Society of America Journal* 49, 127-130.

Sharpley, A.N.; Smith, S. (1994). Wheat tillage and water quality in the Southern Plains. *Soil and Tillage Research* 30, 33-48.

Shi, Z.; Wang, K.; Bailey, J.; Jordan, C.; Higgins, A. (2000). Sampling strategies for mapping soil phosphorus and soil potassium distributions in cool temperate grassland. *Precision Agriculture* 2, 347-357.

Statistics Canada (2016). *Farm and Farm Operator Data; Statistics Canada: Ottawa,* ON, Canada.

Stroia, C.; Morel, C.; Jouany, C. (2007). Dynamics of diffusive soil phosphorus in two grassland experiments determined both in field and laboratory conditions. *Agriculture, Ecosystems and Environment* 119, 6074.

Stroia, C.; Morel, C.; Jouany, C. (2011). Nitrogen fertilization effects on grassland soil acidification: Consequences on diffusive phosphorus ions. *Soil Science Society of America Journal* 75, 112-120

Sun, W.-X.; Huang, B.; Qu, M.-K.; Tian, K.; Yao, L.-P.; Fu, M.-M.; Yin, L.-P. (2015). Effect of Farming Practices on the Variability of Phosphorus Status in Intensively Managed Soils. *Pedosphere* 25, 438-449.

Tran, T.; Giroux, M.; Guilbeault, J.; Audesse, P. (1990). Evaluation of Mehlich-III extractant to estimate the available P in Quebec soils. *Communications in Soil Science and Plant Analysis* 21, 1-28.

Vasu, D.; Singh, S.; Sahu, N.; Tiwary, P.; Chandran, P.; Duraisami, V.; Ramamurthy, V.; Lalitha, M.; Kalaiselvi, B. (2017). Assessment of spatial variability of soil properties using geospatial techniques for farm level nutrient management. *Soil and Tillage Research* 169, 25-34.

Wang, H.-J.; Huang, B.; Shi, X.-Z.; Darilek, J.L.; Yu, D.-S.; Sun, W.-X.; Zhao, Y.-C.; Chang, Q.; Öborn, I. (2008). Major nutrient balances in small-scale vegetable farming systems in peri-urban areas in China. *Nutrient Cycling in Agroecosystems* 81, 203-218.

Watson, C.; Jordan, C.; Kilpatrick, D.; McCarney, B.; Stewart, R. (2007). Impact of grazed grassland management on total N accumulation in soil receiving different levels of N inputs. *Soil Use and Management 23*, 121-128.

Watson, C.; Matthews, D. (2008). A 10-year study of phosphorus balances and the impact of grazed grassland on total P redistribution within the soil profile. *European Journal of Soil Science 59*, 1171-1176.

Webster, R.; Oliver, M.A. (2007). *Geostatistics for Environmental Scientists, 2nd edition; John Wiley & Sons:* Chichester, UK.

West, C.; Mallarino, A.; Wedin, W.; Marx, D. (1989). Spatial variability of soil chemical properties in grazed pastures. *Soil Science Society of America Journal 53*, 784-789.

Whelan, B.; McBratney, A. (2000). The "null hypothesis" of precision agriculture management. *Precision Agriculture 2*, 265-279.

Wilson, H.F.; Satchithanantham, S.; Moulin, A.P.; Glenn, A.J. (2016). Soil phosphorus spatial variability due to landform, tillage, and input management: A case study of small watersheds in southwestern Manitoba. *Geoderma* 280, 14-21.

Wu, C.; Wu, J.; Luo, Y.; Zhang, L.; DeGloria, S.D. (2009). Spatial prediction of soil organic matter content using cokriging with remotely sensed data. *Soil Science Society of America Journal 73*, 1202-1208.

Yuan, Y.; Locke, M.A.; Gaston, L.A. (2009). Tillage effects on soil properties and spatial variability in two Mississippi Delta watersheds. *Soil Science 174*, 385-394.

Zebarth, B.; Drury, C.; Tremblay, N.; Cambouris, A. (2009). Opportunities for improved fertilizer nitrogen management in production of arable crops in eastern Canada: A review. *Canadian Journal of Soil Science 89*, 113-132.

Zhou, Q.; Daryanto, S.; Xin, Z.; Liu, Z.; Liu, M.; Cui, X.; Wang, L. (2017). Soil phosphorus budget in global grasslands and implications for management. *Journal of Arid Environments 144*, 224-235.

Ziadi, N.; Tran, T. (2008). Mehlich 3-Extractable Elements. In *Soil Sampling and Methods of Analysis; Carter, R., Gregorich, E.G., Eds; Taylor & Francis: Boca Raton,* FL, USA; pp. 81-88.

2.11. Tables and figures

Table 2-1. Descriptive statistics of the soil chemical properties for two soil layers (0-5 cm and 5-20 cm) in the young and old grassland fields.

	Unit	0-5 cm						5-20 cm					CV % (%)
		n	Mean	Min	Max	STD	CV $\%^1$	n	Mean	Min	Max	STD	
Young grassland													
Soil pH$_{wa}$ter		151	6.7	5.8	7.4	0.4	5	151	6.8	5.9	7.5	0.4	6
Total carbon	8 kg ^{1}	151	14	5	28	5	38	151	14	5	28	6	40
P$_{M3}$	mg kg~^{1}	151	52	10	161	33	63	151	55	10	203	36	65
Al$_{M3}$	mg kg~^{1}	151	1875	1153	2252	196	10	151	1971	1360	2381	215	11
Te$_{M3}$	mg kg~^{1}	151	104	60	196	24	23	151	111	69	226	27	25
Ca$_{M3}$	mg kg~^{1} <⅛)	151	1180	390	3156	497	42	149	1532	483	3395	605	39
(P/Al)$_{M3}$ ^{2}		151	3.0	0.5	13	0.02	79	151	3.0	0.5	15	0.02	81
Old grassland													
Soil pH$_{wa}$ter		149	5.5	4.7	6.4	0.4	6.5	149	5.3	4.7	6.1	0.3	6
Total carbon	8 kg ^{4}	149	41	11	61	9	23	149	30	10	47	6	20
P$_{M3}$	mg kg~^{1}	149	125	29	327	58	46	149	75	8	226	45	60

| | | 149 | 1831 | 1494 | 2225 | 158 | 9 | 149 | 1891 | 1540 | 2186 | 121 | 6 |

AlM3	mg kg~^{1}	149	1831	1494	2225	158	9	149	1891	1540	2186	121	6
FCM3	mg kg~^{1}	149	187	107	355	32	17	149	169	94	360	36	21
CaM3	mg kg~^{1}	149	1101	325	2457	413	37	149	583	155	1472	263	45
(P/Al)M3		149	7.0	1.4	21	0.04	52	149	4.0	0.4	13	0.03	64

[1] CV: Coefficient Ofvariation;

[2] (P/AI)M3 = (PM3/AlM3) ×100. This soil property is a ratio, determined by soil phosphorus and aluminium concentrations, measured using Mehlich-3 solution.

Table 2-2. Geostatistical parameters of the soil chemical properties for two soil layers (0-5 cm) and (5- 20 cm) from the young and old grassland fields.

Model[1]		Sill ratio[2]		0-5 cm Range R cv[3] [24]		5-20 cm Sill ratio Range R cv[2]			
(%)	(m)	(%)	(m)						
Young grassland									
Soil^P Hwater	Sph	65		81	0.54	Sph	74	62	0.54
Total carbon	Sph		100	29	0.43	Sph	98	28	0.34
PM3	Sph	63		64	0.31	Sph	70	67	0.40
AlM3	Sph	76		40	0.32	Sph	86	40	0.38
FeM3	Sph	58		64	0.31	Sph	69	39	0.39
CaM3	PN	-		-	-	PN	-	-	-
(P/Al)M3	Sph	69		62	0.32	Sph	71	66	0.35
Old grassland									
Soil^P Hwater	Sph	41		37	0.15	Sph	42	39	0.22
Total carbon	Sph	68		31	0.07	Sph	62	62	0.23
PM3	Sph	45		51	0.36	Sph	51	67	0.21
AlM3	PN	-		-	-	Sph	26	32	0.10
FeM3	Sph	69		27	0.32	Sph	43	28	0.26
CaM3	PN	-		-	-	Sph	48	20	0.09
(P/Al)M3	Sph	31		47	0.32	Sph	52	76	0.20

[1] Semivariogram model: Sph: spherical, PN: pure nugget;[2] Sill ratio (%) = [C/(co+c)] x100; this ratio measures spatial dependence or structure according to Whelan and McBratney (2000);[3] Distance at which a semivariance becomes constant;[4] Coefficient of determination of cross-validation.

Table 2-3. Pearson correlation coefficients (r) of the soil available P indices (PM3, and (P/Al)M3 in relationship with soil chemical properties for two soil layers (0-5 cm) and (5-20 cm) from the young and old grassland fields.

	0-5 cm		5--20 cm	
	PM3	(PZAl)M3	PM3	(PZAl)M3
Young grassland				
Soil^P Hwater	0.38 ***	0.37 ***	0.48 ***	0.44 ***
Total carbon	-0.67 ***	-0.64 ***	-0.67 ***	-0.63 ***
AlM3	-0.86 ***	na	-0.87 ***	na
FeM3	0.04 ns	0.10 ns	-0.01 ns	0.02 ns
CaM3	-0.29 **	-0.28 **	-0.13 ns	-0.13 ns
Old grassland				
Soil^P Hwater	0.43 ***	0.46 ***	0.26 **	0.28 **
Total carbon	0.02 ns	0.06 ns	-0.08 ns	-0.09 ns
AlM3	-0.68 ***	na	-0.64 ***	na
FeM3	0.26 **	0.23 **	0.39 ***	0.41 ***
CaM3	0.40 ***	0.43 ***	0.44 ***	0.46 ***

Significance of correlation indicated by *, **, ***, and ns are equivalent to p-value <0.05, <0.01, <0.001, and non-significant respectively; na: non-available

Table 2-4. Multiple regression equations calculated for soil P indices for two soil layers (0-5 cm and 5-20 cm) for the young and old grassland fields.

0-5 cm	5-20 cm
oung grassland	

P_{M3}=-0.135[1] A1M3-0.52[2] TC + 9.513[3] soil pH$_{water}$ +248.6 (R^2 =0.818***)	P_{M3} =-0.123A1M3-1.049 TC +0.773 soil pH$_{water}$ + 307.5 (R^2 =0.789***)
(P/Al) | $_3$ =-2.573 * I ()[1] AI$_{313}$ -O.O()3 TC+ 0.022 soilpH$_{water}$ +0.8 (R^2 =0.893***) (P/A1)$_{M3}$ =-2. 156*IO[1] AI$_{M3}$ - O.0025 TC+0.011 pH$_{water}$ + 0.96 (R^2 =0.868***)	

Old grassland

P_{M3} =-0.227 A1$_{M3}$ +0.377 Fe$_{M3}$[4] -0.015 Ca$_{M3}$[5] + 16.792 SoilpH$_{water}$ + 391.9 (R^2 =0.532***) (P/A1)M3=-3.2*10-[4] Al$_{M3}$ +4.1*IO-[5] FeM$_3$ - 1.3*10-[5] Ca$_{M3}$ + 0.012 soil $_{pHwater}$ +0.99 (R^2 =0.659***)	Pм3-0.235 AIм$_3$ +0.23 FС$_{M3}$ + 0.011 Caм3-4.662 soil pH t$_{waer}$ +499.97 (R^2 =0.547***) (P/A1)м$_3$ =-4.3*10-[4] A1м$_3$ -3.3*10- F6м +2*10-[4]$_3$[5] Ca$_{M3}$ -0.008 soil pH$_{water}$ + 1.2 (R^2 =0.629***)

Significance of regression indicated by *, **, ***, and ns are equivalent to p-value <0.05, p <0.01, p <0.001, and non-significant respectively.

[1]A1м3: Aluminum extracted using Mehlich-3 solution;[2] TC: Total carbon was analyzed with an Elementar Vario MAX CN analyzer;[3] soil pHw½r (1:1 water) was analyzed according to the method of Hendershot et al. [27]; Teм3: Iron extracted using Mehlich-3 solution;[5] Caм3: Calcium extracted using Mehlich-3 solution.

Figure 2-2. Location and sampling strategies design of the studied grassland sites (Google Maps. Accessed 15 June 2020).

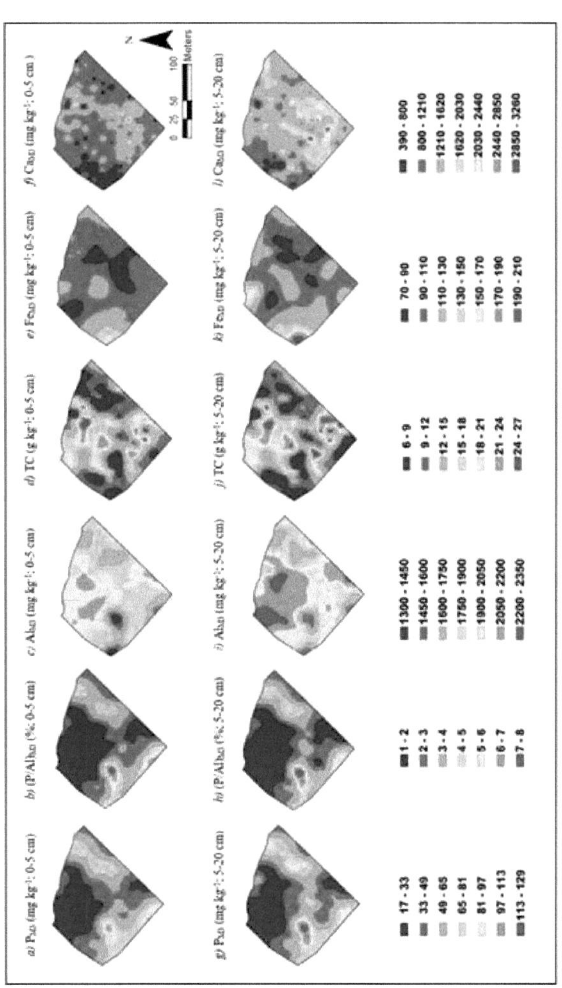

Figure 2-2. Spatial distribution maps of Pмз (a, g), (P/A1)мз (b, h), A1мз (c, I),TC (d, j), Peмз (e, k), and Cамз (f, I) for the 0-5 cm and 5-20 cm soil layers, respectively, from the young grassland field in 2013.

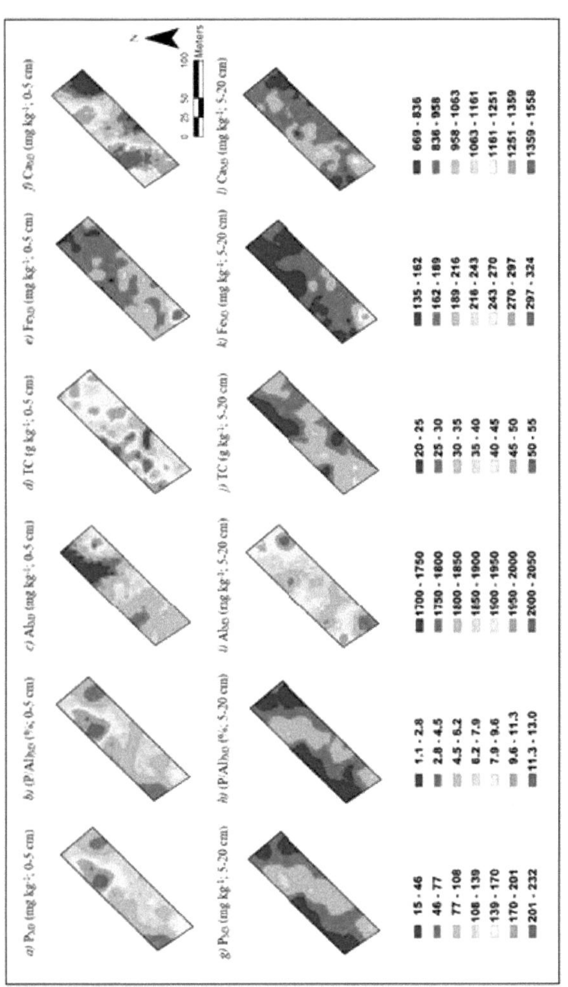

Figure 2-3. Spatial distribution maps of P$_M$з (a, g), (P/A1)мз (b, h), A1$_M$з (c, i),TC (d, j), Fe $\mid$ ᵥ $\mid$ 3 (θ, κ) and Ca$_M$з(f, l) for the 0-5 cm and 5-20 cm soil layers, respectively, from the old grassland field in 2013

Tillage Management Impacts on Soil Phosphorus Variability under Maize-Soybean Rotation in Eastern Canada

Jeff D. Nze Memiaghe[1,2] , Athyna N. Cambouris[1,] , Noura Ziadi[1] and Antoine Karam[2]*

[1]Agriculture and Agri-Food Canada, Quebec Research and Development Centre, 2560 Hochelaga Boulevard, Québec City, QC G1V 2J3, Canada;

[2]Soils and Agri-Food Engineering Department, Université Laval, 2425 rue de l'Agriculture, Québec City, QC G1V 0A6,

*Published in Soil Systems MDPI2022, 6(2), 45; **https://doi.org/10.3390/soilsystems6020045***

3.1. Core Ideas

•   Soil P accumulation was observed under the NT system compared to the CT system, presenting a greater risk of P pollution.

•   Relationships between soil-available P indices and other chemical properties differed between the contrasting tillage practices.

•   The variability of soil P in both fields ranged from moderate to very high (32-60 %), indicating that the uniform P fertilizer recommendations currently used for maize-soybean rotations are not suitable for large (10 ha) fields.

•   The use of kriged maps demonstrated the importance of applying P at the right rate in the right location to develop accurate future P recommendations.

Keywords: soil P indices; spatial variability; no-tillage; geostatistics; P recommendations; precision agriculture.

3.2. Summary

Conservation tillage, including direct seeding (D.S.), is increasingly being used in place of conventional tillage (CT) to reduce soil erosion, improve water conservation and prevent land degradation. However, the practice of SD increases soil phosphorus (P) stratification, causing P runoff and eutrophication. For sustainable P management, fertilisation must be balanced between P sources and actual crop demand. It is therefore important to better understand the spatial variability of P in fields under SD in order to reduce P losses to the environment. Very little information exists on the impact of tillage on the spatial variability of P at field scale in precision agriculture. This study examines the impacts of tillage on the spatial variability of available P in two commercial fields under corn-soybean rotation, rated TC (10.8 ha) and SD (9.5 ha), for the purpose of improving phosphate fertilizer recommendations in eastern Canada. NPK fertilizers were applied to the soils (Humic Gleysols) according to local recommendations. Soil samples were taken in autumn 2014 using regular 35 m by 35 m grids, at depths of 05 and 5-20 cm, providing for fields under TC and SD, 141 and 134 geo-referenced points, respectively. Available P and other elements were analysed by Mehlich-3 (M3) extraction and the P saturation index $(P/Al)_{M3}$ was calculated. The variability of available soil P in both fields ranged from moderate to very high (32% to 60%). An average $(P/Al)_{M3}$ of 3% was measured in both layers under CT, compared with 8% in the 0-5 cm layer and 6% in the 5-20 cm layer under SD. The relationships between P indices and other elements differ according to tillage practices. This study highlights the importance of improving phosphate fertilizer recommendations in eastern Canada.

Keywords: soil P indices; spatial variability; no-till; geostatistics; P recommendations; precision agriculture

3.3. Abstract

Conservation tillage, including no-tillage (NT), is being used increasingly with respect to conventional tillage (CT) to mitigate soil erosion, improve water conservation and prevent land degradation. However, NT increases soil phosphorus (P) stratification, causing P runoff and eutrophication. For sustainable P management, fertilization must be balanced between P sources and actual crop demand. To reduce P losses to the environment, it is important to better understand P spatial variability in NT fields. Little is known about tillage impacts on field-scale P spatial variability in precision agriculture. This study examines tillage impacts on spatial variability of soil-available P in a maize-soybean rotation, in two commercial fields, denoted CT (10.8 ha) and NT (9.5 ha), with the aim of improving P fertilizer recommendations in Eastern Canada. NPK fertilizers were applied to the soils (Humic Gleysols) following local recommendations. Soil samples were collected in fall 2014 in regular 35 m by 35 m grids, at 0-5 and 5-20 cm depths, providing 141 and 134 georeferenced points for CT and NT fields, respectively. Available P and other elements were analysed by Mehlich-3 extraction (M3), and the P saturation index $(P/Al)_{M3}$ was calculated. Variability of soil available P in both fields ranged from moderate to very high (32% to 60%). A mean $(P/Al)_{M3}$ of 3% was found in both layers under CT, compared to 8% in the 0-5 cm layer and 6% in the 5-20 cm layer under NT. Relationships between P indices and other elements differed between tillage practices. This study highlights the need to improve P fertilizer recommendations in Eastern Canada.

Keywords: soil P indices; spatial variability; no-tillage; geostatistics; P recommendations; precision agriculture

3.4. Introduction

Maize (*Zea mays* L.) and soybean (*Glycine max* L.) are dominant crops in Eastern Canada, representing 75% of cultivated land and 90% of agricultural production (MAPAQ, 2020). From 2014 to 2018, 5.5 million tons of these crops were produced, generating total revenue of US $2 billion for the province of Quebec (MAPAQ, 2020). Owing to their high economic value, these crops are cultivated by local farmers using a variety of tillage methods, including conventional tillage (CT) and conservation tillage methods such as reduced tillage and no-tillage (NT). CT, which involves preparing the soil with a moldboard plow or a disc plow (Malvezi et al., 2019), improves maize productivity by enhancing its root biomass (Guan et al., 2014) and grain yield (Liu and Wiatrak, 2012). However, this tillage method deteriorates the soil structure and increases the risk of soil erosion, which can lead to higher soil nutrient losses in crops fields (Alvear et al., 2005; Li, 2017). To address these problems, conservation tillage practices including NT have been introduced with the broader aim of promoting economically sustainable farming systems (Kassam et al., 2012; Soane et al., 2012).

Conservation tillage methods, characterized by minimal soil mixing and disturbance, have been adopted increasingly in recent years, owing to their environmental benefits, such as reduced water and wind erosion (Olson and Ebelhar, 2009). The NT method improves soil physical properties by reducing compaction, soil erosion, runoff and drainage risk (Dick, 1992; Blevins and Frye, 1993; Andraski et al., 2003; Daverede et al., 2003; Olson and Ebelhar, 2009; Friedrich et al., 2012). In addition, NT is recognised as the best agronomical practice for maintaining soil integrity in intensively cultivated and large agricultural fields worldwide (Blevins and Frye, 1993; Friedrich et al., 2012; De Santiago et al., 2019). For instance, it is estimated that approximately 9% (125 million ha) of global arable land was under NT management in 2011 (Friedrich et al., 2012; Pittelkow et al., 2015). Canadian farmland under NT management covered an area of 19.5 million ha in 2016 (Statistics Canada, 2016). In Quebec, farmland area under NT has doubled to more than 69% (Statistics Canada, 2016), largely due to local government subsidies paid to farmers for using NT practices between 2009 and 2013. Further increases in the use of NT have been observed in recent years (Statistics Canada, 2016), owing to the benefits from the standpoint of agronomic (Lafond et al., 2011) and economic performances (Holm et al., 2006).

Agri-environmental concerns have nonetheless raised general concerns about no-tillage agronomic practices, in relation to P stratification (Cade-Menum et al., 2010; Messiga et al., 2010; Abdi et al., 2014), transport and runoff to surface waters (Puustinen et al., 2005; Allaire et al., 2011), which may cause P eutrophication in rivers (Sun et al., 2015). Sustainable management of P involves balancing fertilizer application between P supply (sources) and actual crop demand for soil P (Tunney, 1990). To achieve this objective, it is important to increase our knowledge of the spatial variability of soil P under NT management for developing more rational and cost-effective P fertilization programs that will enhance yields and reduce P losses to the environment. The spatial variability of soil P can be controlled through two main agronomic strategies: (1) variable rate applications (VRA) and (2) the use of management zones (MZs) in precision agriculture.

Many studies (Mallarino, 1996; Borges and Mallarino, 1998; Fernandez and Schaefer, 2012; Dalchiavon et al., 2017; Malvezi et al., 2019) have examined the spatial variability of soil P under various tillage systems at different sampling, plot or field scales. Cambouris et al. (2017) characterized the distribution of Mehlich-3 P (P_{M3}) concentrations and agri-environmental P saturation $(P/Al)_{M3}$ at decimetric scale in NT and moldboard plow (MP) plots in a long-term maize-soybean rotation (>20 yr). They observed high CVs associated with P_{M3} data in both MP (77% and 63% at 0-5 and 5-20 cm, respectively) and NT plots (46% and 66% at 0-5 and 5-20 cm, respectively). Nevertheless, these authors found that the 2D geospatial model related to tillage was not detected by the nested sampling grid used at plot-scale in their study. They suggested the use of an appropriate geostatistical sampling strategy at field scale. Sun et al (2015) reported that studies on the spatial variability of P at plot or experimental scales allow limited interpretation over a larger field area.

In Eastern Canada, local P recommendations have been developed for uniform P agronomic applications (CRAAQ, 2010), based mainly on both soil P availability [(P_{M3}) and $(P/Al)_{M3}$] index, regardless of field size, soil P variability or their crop productivity potential. Soil variability and associated potential yield differences in a given field are responsible for the development of large field-scale heterogeneity in the distribution pattern of soil nutrients including P (Nolin et al., 1999; Nze Memiaghe et al., 2021). P fertilization based on uniform recommendations may result in over-fertilization in large field areas and also create underfertilized areas (Cambouris et al., 1999). In Quebec, the critical environmental threshold for soil P corresponds to a $(P/Al)_{M3}$ value of 8% for fine-textured soils (Pellerin et al.,2006; MDDELCC, 2017). Consequently, a better understanding of the spatial variability of soil P in large fields will help to establish accurate P recommendations and ensure that P is applied at the right rate in the right locations.

To our knowledge, no studies have investigated the impacts of contrasting soil tillage systems (CT vs. NT) on the spatial variability of soil P in commercial fields in Eastern Canada under maize-soybean rotation, the dominant legume-cereal association in North America. The main goal of this study was to evaluate the spatial variability of soil-available P under two contrasted tillage systems for developing accurate future P fertilizer recommendations in Eastern Canada. The specific objectives were: to (1) investigate field-scale spatial variability of soil P indices-referred to as P_{M3} and $(P/Al)_{M3\text{-AND}}$ other selected soil chemical properties (soil pH_{H2O}, total carbon, Al_{M3}, Ca_{M3}, Fe_{M3}) by using descriptive statistics and geostatistical tools; to (2) study the relationships between soil P indices and these soil chemical properties based on Spearman correlations and multiple regression equations; and to (3) evaluate the current P fertilizer recommendations using prescription maps of kriged values of $(P/Al)_{M3}$. The knowledge gained from this research on the effects of tillage management on the spatial variability of soil P indices will provide significant information for optimizing P fertilization in sustainable agriculture systems.

3.5. Materials and methods

3.5.1. Site Description and Soil Sampling

The study was conducted near Montreal, in the Montérégie region of the province of Quebec, Canada. The Montérégie region is located in the physiographic region of the St. Lawrence Lowlands, which is characterized by very flat landforms (0-5%) (Wang et al., 1995). Two commercial fields were selected to represent two contrasting soil tillage practices. One field, denoted the CT field, located at St. Marc-sur-Richelieu, was managed under CT (45° 42' N; 73°14' W; 10.8 ha, established in 1994), and the second field, denoted the NT field, located at La Presentation, was managed under NT (45° 36' N; 73°02' W; 9.5 ha, established in 1994). Both fields were managed under maize-soybean rotation.

In the CT field, tillage has consisted of one plowing operation (20 cm deep) since 2009 using a moldboard plow in the fall after harvest, followed by disking and harrowing (10 cm deep) each spring before seeding. In the NT field, flat direct seeding was performed from 1994 to 2014 onward, with crop residues (30% of residues) left on the ground after harvest. NPK fertilizer sources were as follows: N fertilizer provided from ammonium nitrate (34-0-0); P fertilizer provided from triple superphosphate (0-46-0); and K fertilizer provided from muriate of potash (0-0-60). For each crop, application rates were applied based on previous soil chemical analysis and following province of Quebec recommendations (CRAAQ, 2010). Both fields were seeded with maize in May 2014 and harvested in late October 2014. At seeding, the NPK fertilizers were band-applied similarly in both fields under the two tillage systems. Using commercial fertilizers, P and K were both applied at the seeding stage, while N was split at the seeding stage, and at the 6-8 leaf stage, according to local fertilizer recommendations (CRAAQ, 2010). For maize, the seeding rows were spaced 75 cm apart in both fields. For soybean, the seeding rows were spaced 30 cm apart in both fields. Crop management, maize and soybean fertilization practices were based on local recommendations for the province of Quebec (CRAAQ, 2010).

Both soils were classified as Orthic Humic Gleysols and presented similar soil series (St-Urbain and Kierkoski) (Nolin and Lamontagne, 1990). Soil textures for both fields varied from clay to clay loam. They were classified to the same soil texture group (i.e., fine-textured soils with >30% clay) used in the province of Quebec recommendations (CRAAQ, 2010). Both fields were poorly drained. Meteorological data from the study region were summarized (Table 3-1). The mean annual temperature is 6°C and total precipitation is 973 mm at St. Marc-sur-Richelieu. The mean annual temperature is 5.9°C and total precipitation is 984 mm at La Presentation.

An intensive soil sampling was performed on November 2014 using a grid design with a sampling interval of 35 m × 35 m, providing 141 and 134 georeferenced sampling points for the CT and NT fields, respectively (Figure 3-1). A composite soil sample made up of four cores was taken within a 1 m radius of each sampling point at two soil depths (0-5 cm and 5-20 cm) using a 0.05 m diameter Dutch auger. A total of 282 and 268 soil samples were collected from the CT and NT fields, respectively. At both locations, soil samples were collected after the corn harvest.

Soil samples were air-dried, ground and sieved through a 2 mm sieve for soil characterization. Soil pH_{H2O} (1:1 water) was measured according to Hendershot et al (2008). The soils were extracted using Mehlich- 3 solution at a 1:10 soil: solution ratio (Ziadi and Tran, 2008), and the concentrations (mg kg^{-1}) of available phosphorus (P_{M3}), iron (Fe_{M3}), calcium (Ca_{M3}) and aluminum (Al_{M3}) were determined by inductively coupled plasma optical emission spectroscopy (ICP-OES; Model, 4300DV, PerkinElmer, Inc., Shelton, CT, USA). The soil P agri-environmental indicator $(P/Al)_{M3}$, which is a ratio (%), was calculated from the soil P_{M3} and Al_{M3} concentrations. Total carbon (TC, g kg^{-1}) content was determined using an Elementar Vario MAX CN analyzer (Elementar 137 Analysensyteme GmbH, Hanau, Germany).

Two topographic parameters were produced for each field using the digital terrain model derived from light detection and ranging (LiDAR) imagery of the province of Quebec. These parameters are elevation and the topographic wetness index (TWI, where TWI = ln(αZtan(β)) is as described by Beven and Kirkby (1979) and Sorensen et al. (2006). The value of TWI depends on α, which represents the upslope area per unit contour length, and β, where tan(β) is the local slope. The spatial resolution of these rasters was 1 m. The CT and NT fields were extracted from the respective LiDAR sheets 31H11 NE and 31H11 SE, using the" extract by mask" ArcGIS tool (ArcGIS Software version 10.3 [ESRI, Redlands, CA, USA]). Mean elevation and TWI values were generated from each sampling point using the buffer method within 1 m radius and the" Zonal Statistic as Table" ArcGIS tool (ArcGIS Software version 10.3 [ESRI, Redlands, CA, USA]).

3.5.2. Statistical and Geostatistical Analyses

Descriptive statistics were analysed using SAS software version 9.4 (SAS Institute Inc., Cary, NC, USA, (SAS Institute, 2010) to investigate soil pH_{H2O}, P_{M3}, Al_{M3}, $(P/Al)_{M3}$, Fe_{M3}, Ca_{M3}, TC, elevation and TWI and their intensity of variability. The minimum and maximum values, the mean, CV and standard deviation of the mean (SDM) were determined for the 0-5 cm and 5-20 cm soil layers from each field. Data were summarized using SigmaPlot software version 14.5 (Systat Software Inc., Palo Alto, CA, USA). The intensity of the variability of the different soil properties was classified based on the approach of Nolin and Caillier (1992) using five classes based on the CV: (1) low (CV < 15%); (2) moderate (15% < CV < 35%); (3) high (35% < CV < 50%); (4) very high (50% < CV <100%); and (5) extremely high (CV > 100%). Spearman correlation coefficients (r) between soil P indices [P_{M3}, $(P/Al)_{M3}$ and other soil chemical properties (Al_{M3}, Ca_{M3}, Fe_{M3}, TC and soil pH_{H2O}) were determined for each soil layer using SAS software (SAS Institute, 2010). Spearman coefficients between soil P indices and topography properties (elevation, TWI) were only determined for each topsoil layer. Data were analysed using the CORR and Univariate procedures.

To determine the relation between the response variables [$(P/Al)_{M3}$ and P_{M3}] and the explanatory variables (Al_{M3}, Ca_{M3}, Fe_{M3}, TC and soil pH_{H2O}), a mixed regression model was used with a spherical spatial correlation structure between observations. Since measurements were taken at different spatial points (X,Y)

$^{i i}$ and at two different depths (0-5 and 5-20 cm) at each point, a different spherical spatial correlation   AA

$c‰$

was assumed for the two depths. Letting ^{n} and ^{o} denote the nugget and the range estimates, respectively, the correlation between any two observations a distance d apart, ^{p} $(-d)$ is equal to

... $i(1 - Cn)*(1 -1.5(d / ⅞) + 0.5(d / ⅜)3)$ if $d < ⅜$ p$(d) = <$

0if $d \geq cc_0$

The regression models were fitted to each field separately. For each response variable, the Box-Cox methodology was used to transform the response variable. The explanatory variables were added sequentially to the model, with the interaction depth factor evaluated each time using the Bayesian information criterion (BIC) until a minimum value was reached. Thus, the final model consisted of the most important explanatory variables and their interaction with the depth factor, which allowed an equation model to be derived for each depth. A pseudo R^2 value was calculated to evaluate the performance of the model. All analyses were performed using the MIXED procedure in SAS (SAS Institute Inc., Cary, NC, USA, release 9.4) at the 0.05 level of significance. The normality assumption was validated using the Shapiro-Wilk statistic on the normalized residuals of the model.

Geostatistics provide the basis for interpolating the spatial variability of soil properties (Webster and Oliver, 1990; Pohlmann, 1993; Cambardella et al., 1994). Geostatistical analyses of P_{M3} (mg kg^{-1}), $(P/Al)_{M3}$ (%), Al_{M3} (mg kg^{-1}), TC (g kg^{-1}), Ca_{M3} (mg kg^{-1}), soil pH_{H2O} and Fe_{M3} (mg kg^{-1}) data were performed for each soil layer using the GS+ (version 9) software (Gamma Design Software, LLC., Plainwell, MI, USA) (Robertson, 2008). The spatial dependence or structure of these soil properties was evaluated using isotropic and anisotropic semivariograms. Semivariogram parameters were generated for each theoretical model (spherical, exponential and linear). The corresponding nugget (c_0), partial sill (C), sill (C0 + C) and range values of the best-fitting theoretical model were calculated. The partial sill ratio [C/(C0 + C)] was used to determine the spatial dependence of the soil properties for the CT and NT fields. Semivariograms with a partial sill ratio of <25%, 25% to 40%, 40% to 60%, 60% to 75% or >75% were considered to have a low, low moderate, moderate, strong moderate or strong spatial dependence, respectively (Whelan and Mc Bratney, 2000). The range indicates the maximum distance at which sample points are correlated (Vieira et al., 1983). Spatial variability maps were generated for each soil layer using the block kriging method, with a block size of 1 m × 1 m, and were evaluated using cross-validation analysis (R^2 cv) (Kravchenko et al., 2002).

Prescription maps were produced using the kriged values of soil P indices $(P/Al)_{M3}$ for maize and soybean fields using ArcGIS software version 10.3 (ESRI, Redlands, CA, USA), in order to identify different P agri-environmental classes for Quebec and the corresponding areas in both fields. Lastly, accurate P fertilizer recommendations (kg P_2O_5) were developed for maize and soybean in these two commercial fields. These results were compared to the local uniform P fertilizer recommendations [based on $(P/Al)_{M3}$ mean values] established for Eastern Canada (CRAAQ, 2010).

3.6. Results

3.6.1. Descriptive Statistics of Topography Properties

In the CT field, elevation values ranged from 19.2 to 19.9 m, with a mean value of 19.4 m. The TWI values ranged from 6 to 13 with mean value of 9. The TWI values ranged from 6 to 13 with a mean value of 9. The CVs for elevation (0.8%) and TWI (13%) were low. In the NT field, elevation values ranged from 34.2 to 35.0 m with a mean value of 34.8 m. The TWI values ranged from 7 to 14 with a mean value of 10. The CVs for elevation (0.3%) and TWI (11%) were low. Similar TWI values were obtained, and the CVs for topography properties were low.

3.6.2. Descriptive Statistics of Soil Phosphorus Indices and Other Soil Chemical Properties

In the CT field, similar mean values for the soil P indices [P_{M3}, and $(P/Al)_{M3}$] and other soil chemical properties were obtained for both soil layers (0-5 cm and 5-20 cm), with the exception of Fe_{M3} and Al_{M3} (Figure 3-2). Mean values of Fe_{M3} and Al_{M3} were, respectively, 8% and 12% greater in the 5-20 cm soil layer relative to the 0-5 cm layer. The CVs ranged from 3% to 59% for both soil layers (Figure 3-2). The highest CVs were obtained for soil P indices. The CVs were classified as follows: low for soil pH_{H2O}, TC, Al_{M3}, Fe_{M3} and Ca_{M3}, and high to very high for both soil P indices.

In the NT field, mean values of soil P indices, TC and Ca_{M3} were higher in the 0-5 cm layer than in the 5-20 cm layer. Conversely, the mean values of soil pH_{H2}O, Al_{M3} and Fe_{M3} (Figure 3-2) were higher in the 5-20 cm layer than in the 0-5 cm layer. The CVs ranged from 7% to 43% for both soil layers (Figure 3-2). The highest

CVs were obtained for soil P indices. The CVs were classified as follows: low for soil pH_{H2O}; moderate for TC, Fe_{M3} and Ca_{M3}; low to moderate for Al_{M3}; and moderate to high for the two soil P indices (Figure 3-2). The intensity of variation of the soil P indices was higher in CT than in NT for both soil layers.

3.6.3. Geostatistical Parameters of Soil Phosphorus Indices and Other Soil Chemical Properties

In the CT field, soil P indices and the other soil chemical properties (except for TC and Ca_{M3}) were fitted with spherical and linear models (Table 3-2). Pure nugget models (PNs) were used for TC and Ca_{M3} in both soil layers. Spatial dependence ratios of soil P indices and the other chemical properties ranged from 23% to 85% for both soil layers. Spatial dependence ratios were classified as follows: low moderate for soil pH_{H2O}; moderate to strong for Al_{M3}; low moderate for Fe_{M3}; and low to low moderate for soil P indices (Table 3-2). Spatial ranges varied between 46 and 183 m, extending beyond the 35 m × 35 m sampling grid. Cross-validation coefficients (R^2 cv) ranged from 0.01 to 0.69 for soil chemical properties and from 0.15 to 0.17 for soil P indices.

In the NT field, soil P indices and the other soil chemical properties were fitted with spherical and exponential models (Table 3-2). Spatial dependence ratios of soil P indices and the other soil chemical properties ranged from 33% to 99%. They were classified as follows: strong for soil pH_{H2O}; moderate for Al_{M3}; moderate to strong for TC and Ca_{M3}; and low moderate to moderate for soil P indices (Table 2). Spatial ranges varied from 55 to 77 m, extending beyond the 35 m × 35 m sampling grid. R^2 cv ranged from 0.07 to 0.36 for soil chemical properties and from 0.06 to 0.17 for soil P indices. The spatial dependence of soil P was higher for both NT soil layers compared to CT soil layers. Conversely, R^2 cv values were generally lower for the NT field relative to the CT field. Higher R^2 cv values (R^2 cv > 0.6) indicate a good fit for kriged map reliability. Consequently, the R^2 cv values for the CT field indicate a relative fit for map reliabilities for most soil chemical properties, including soil P indices, while a weaker fit was found for map reliabilities of soil chemical properties in the NT field.

3.6.4. Spatial Distribution of Soil Phosphorus Indices and Other Soil Chemical Properties

In the CT field, visual similarities were observed between P_{M3} and $(P/Al)_{M3}$ spatial distribution maps. Spatial distribution values of soil P indices were similar in both soil layers (Figure 3-3a,b,g,h). Spatial distribution values of TC were higher in the 0-5 cm soil layer compared to the 5-20 cm soil layer (Figure 3-3d,j). Spatial distribution values of Al_{M3} and Fe_{M3} were lower in the 0-5 cm soil layer compared to the 5-20 cm soil layer (Figure 3-3c,e,i,k). In the NT field, visual similarities were also observed between P_{M3} and $(P/Al)_{M3}$ spatial distribution maps (Figure 3-4a,b,g,h). However, spatial distribution values of soil P indices, TC and Ca_{M3} were higher in the 0-5 cm soil layer than in the 5-20 cm soil layer (Figure 3-4a,b,d,f-h,j,l). Spatial distribution values of these soil chemical properties ranged from 35 to 19 mg kg^{-1}, 5.0% to 12.5%, 19 to 27 g kg^{-1} and 1600 to 4540 mg kg^{-1} compared to 35 to 95 mg kg^{-1}, 2.0% to 9.5%, 13 to 25 g kg^{-1}, 1600 to 2860 mg kg^{-1} in the 0-5 cm and 5-20 cm NT soil layers, respectively. The spatial distribution value of Al_{M3} was similar in both soil layers (Figure 3-4c,i). The spatial distribution value of Fe_{M3} was lower in the 0-5 cm soil layer than in the 5 20 cm soil layer (Figure 4e,k).

Spatial distribution maps also revealed visual associations between soil P indices and other soil elements. In the CT field, high soil P indices values correspond to the high TC and Ca_{M3} concentrations in both soil layers (Figure 3-3a,b vs. Figure 3-3d,f; and Figure 3-3g,h vs. Figure 3-3j,l). In the NT field, the soil P accumulation regions in the soil layer (0-5 cm) correspond to high Ca_{M3} concentrations, providing evidence of P_{M3}-Ca_{M3} accumulation (Figure 3-4a,b,f). The highest soil P accumulation regions in the soil layer (0-5 cm) correspond to the lowest Fe_{M3} concentrations.

3.6.5. Relationships between Soil Phosphorus Indices and Other Soil Chemical Properties

In the CT field, the Spearman correlation coefficients with the highest significant relationships were observed between soil P indices and soil TC, Fe_{M3} and Ca_{M3} in both soil layers (Table 3-3). In the NT field, the Spearman correlation coefficients with the highest significant relationships were observed between soil P indices and soil TC, pH_{water} and Ca_{M3}. In addition, a significant relationship was observed between soil P indices and Fe_{M3} in the 5-20 cm soil layer in the NT field (Table 3-3).

Spatial multiple regression equations were generated for the soil P indices in order to investigate the influence of selected soil chemical properties on the variability of soil P indices (Table 3-4). In the CT field, 45% to 72% of the variability of soil P indices was explained significantly mainly by TC, Fe_{M3}, Al_{M3} and Ca_{M3}. In the NT field, 28% to 39% of variability of soil P indices was only explained significantly by Ca_{M3} and Fe_{M3}.

3.6.6. Phosphorus Fertilizer Recommendations Based on Prescription Maps

In the CT field, kriged spatial values for soil $(P/Al)_{M3}$ ranging from 1% to 4.5% correspond to two Quebec agronomic classes (0% to 2.5%, and 2.6% to 5%) for maize and soybean. According to these kriged values for $(PZAl)_{Ms}$, the P fertilizer recommendations should be 60 or 80 kg P_2O5 ha^{-1} for maize and 20 or 60 kg P_2O5 ha^{-1} for soybean (Figure 3-5a,e). The applied areas (A1 and A2) from these kriged values were 6.1 and 4.7 ha. Thus, in the CT field, accurate total P recommendations were 770 and 460 kg P2O5 for maize and soybean, respectively (Table 3-5). Based on the average value of $(P/Al)_{M3}$ in the CT field (2.7%), the uniform P agronomic recommendations should be 60 kg P2O5 ha^{-} 1, for a total of 648 kg P2O5 for maize, and 20 kg P2O5 ha^{-1} , for a total of 216 kg P2O5 for soybean (Table 3-5).

In the NT field, kriged spatial values for soil $(P/Al)_3$, which ranged from 2% to 12.5%, correspond to three Quebec agronomic classes, i.e., 2.6% to 5%, 5.1% to 10.0% and 10.1% to 15% for maize, and 2.6% to 5%, 5% to 7.5% and 7.6% to 15% for soybean. The P fertilizer recommendations should be 20, 40 or 60 kg P2O5 ha^{-1} for maize, according to these $(P/A1)_{M3}$ kriged values (Figure 3-5b). The applied areas corresponding to these kriged values were 0.7, 8.8, 6.9 and 2.6 ha (Table 3-5). For soybeans, the P fertilizer recommendations should be 20 or 0 kg P2O5 ha^{-1} , according to these $(P/A1)_{M3}$ kriged values (Figure 3-5f). The applied areas corresponding to these kriged values were 2.6, 3 and 6.5 ha (Table 3-5). In the NT field, the accurate total P recommendations were 366 and 0 kg P2O5 for maize and soybean, respectively (Table 3-5). Based on the average value of $(P/Al)_{M3}$ in the N I field (7.9%), the uniform P agronomic recommendations should be 40 kg P2O5 ha^{-1} , for a total of 380 kg P2O5 for maize, and 0 kg P2O5 ha^{-1} , for a total of 0 kg P2O5 for soybean (Table 3-5).

3.7. Discussion

3.7.1. Effects of Topography Properties on Phosphorus Transport

Both fields used in this study were flat, as confirmed by previous soil studies conducted in the region (Wang et al., 1995; Nolin et al., 1999). The TWI is a factor that promotes P transport in temperate agricultural landscapes (Heathwaite et al., 2005; Roberts et al., 2017). Mean TWI values were higher than in Li et al. (2020). A higher TWI is associated with increased surface runoff, which may result in greater export of nutrients (Li et al., 2020), such as P. However, the CV values of TWI were lower than in other studies (Li et al., 2020; Salekin et al., 2021). Thus, similar TWI values from the contrasting tillage fields and their respective low CVs provide evidence of failure to take account of these topographic properties in relation to P transport and spatial variability. This is confirmed by results showing a lack of significant relationships between these topography properties and soil P indices (Table 3).

3.7.2. Effects of Soil Tillage Practices on Soil Phosphorus Stratification

The mean values of soil P indices were higher in the 0-5 cm soil layer relative to the 5-20 cm layer under the NT system. In contrast, the mean values were similar in both soil layers in the CT field. Our results show that P stratification exists (Figure 3-4a,b,g,h). Soil P stratification is defined as high soil P concentrations at the soil surface and rapidly decreasing concentrations with depth (Selles et al., 1999). Many studies (Duiker and Beegle, 2006; Bertol et al., 2007; Cade-Menum et al., 2010; Messiga et al., 2012, Abdi et al., 2014; Rodrigues et al., 2016) have found soil P stratification in agricultural plots managed using the NT system relative to the CT.

In this field-scale study, soil P stratification can be explained by crop residues and the lack of mixing of soil and fertilizers. Previous studies (Duiker and Beegle, 2006; Piegholdt et al., 2013) reported that P stratification in soils under NT has three causes: (1) annual broadcast or band application of P fertilizer at or near the same sowing row; (2) absence of mixing of P fertilizers with soil and crop residues left on the soil surface after harvest; and (3) immediate leaching of P from crop residues.

The results obtained show that increased soil P accumulation is associated with NT, which points to a higher agri-environmental risk for NT field management in Eastern Canada. The mean $(P/Al)_{M3}$ index increased (7.9%) in the 0-5 cm NT soil layer relative to the 5-20 cm NT soil layer (6%). These results also have environmental implications for soil P. The mean $(P/A1)_{M3}$ value for the NT field (7.9%) was close to Quebec's national critical threshold for P saturation for fine-textured soils (8%), indicating that $(P/Al)_{M3}$ values higher than this can result in water contamination. Consequently, NT affects the sustainable management of soil P, increasing the risk of P pollution in Eastern Canada.

3.7.3. Variability of Soil Phosphorus Indices under the Different Soil Tillage Practices

Among all properties measured in both fields, the highest CVs were obtained for soil P indices. These results are in line with other studies (Malvezi et al., 2019; Metwally et al., 2019). In this field-scale experiment, the CVs of soil P indices ranged from moderate to very high (32% to 60%). Many studies (Mallarino, 1996; Cambouris et al., 2017; Dalchiavon et al., 2017) obtained similar results under different tillage systems at different scales. For instance, Malvezi et al. (2019) reported CVs (n = 100; 0-20 cm) for soil-available P ranging from moderate to high (32% to 76%) under two contrasting CT and NT experimental plot-scale systems cultivated for more than three decades.

The CVs of soil P indices were high (32% to 43%) in the NT field. Other studies (Kitchen et al., 1990; Tyler and Howard, 1991; Mallarino, 1996) reported that high CVs of soil P in NT fields were mainly caused by (1) limited mixing of soil, crop residues and fertilizers, and (2) the high residual P levels in soil surface in NT fields. A high CV of soil P in NT fields has implications for developing future agronomic strategies for sustainable management of P based on delineating MZs or VRA for accurate P recommendations in precision agriculture.

The intensity of variation of soil P indices was lower in both soil layers (0-5 cm and 5-20 cm) under the NT system, compared to CT. These results are similar to the findings of Cambouris et al. (2017) in Eastern Canada, and stand in contrast to those of Malvezi et al. (2019) in southern Brazil. Consequently, variability of soil P may be reduced under NT at the scale used in the present study. The lower CV obtained with the NT system will positively impact future soil P sampling strategies, based on a limited number of soil samples.

3.7.4. Geostatistics of Soil Phosphorus Indices under Different Soil Tillage Practices

The spatial dependence of soil P indices was moderate (23% to 46%), indicating that soil P variability in the contrasting tillage fields resulted from interaction between intrinsic (soil types) and extrinsic soil factors, such as soil/crop management practices (tillage mode and application of fertilizers or manure). Moderate spatial dependence (25% < R < 75%) results from interaction between intrinsic and extrinsic soil factors (Cambardella et al., 1994; Vasu et al., 2017). These results are similar to those of Dalchiavon et al. (2017), who found a moderate spatial dependence (54%) for soil P in a 5 ha soybean field under a NT system. In contrast, other studies (Cambouris et al., 2017; Malvezi et al., 2019) reported pure nuggets (PNs) under CT and NT long-term experiments. These PNs were explained by (1) the plot scale of these studies, and (2) a strong overall variability of soil P indices, which overshadowed soil P patterns at the plot-scale used in the study.

Spatial ranges revealed that the sampling grid used to measure spatial variability in both fields was appropriate. The present field-scale study also revealed that the 2D geospatial model related to tillage was detected for soil P, based on the sampling grid used initially, compared to the plot-scale experiment conducted by Cambouris et al. (2017). In this earlier study, the 2D geospatial model related to tillage was not detected by the sampling grid used. Furthermore, the smallest range values, 46 m and 55 m, corresponded to soil P indices from CT and NT fields, respectively (Table 3-2), indicating that these minimum ranges will strongly influence the optimum grid for determining soil P and other chemical properties in both fields (McBratney and Webster, 1983). Consequently, future strategies for sampling soil P and other chemical properties for geostatistical research should be planned according to the tillage systems used, taking into account the range of P values corresponding to each tillage system.

Spatial dependence of soil P was greater under the NT system compared to CT. This can be explained by the reduction in soil tillage activities under NT fields over a long period, which affects the long-term spatial structure of soil P at field-scale. In contrast, the lowest R^2 cv values for soil P indices were observed with stronger spatial structures in the NT field. Based on the hypothesis of Kravchenko (2003), we can infer that the geostatistical models traditionally used are not suited to NT fields. Previously, Mallarino (1996) confirmed this idea in his field-scale study on NT systems in Iowa. Further studies need to be carried out using other adapted models to achieve better prediction of spatial structure of soil P indices in NT fields.

3.7.5. Spatial Distribution Maps and Relationships between Soil Phosphorus Indices and Other Soil Chemical Properties

The R^2 cv is an important reliability factor for producing kriged spatial maps of soil P indices with the aim of preventing P pollution. R^2 cv values (R^2 cv < 0.2) were obtained for most soil properties. These values are similar to those of other studies (Cambouris et al., 2017; Nze Memiaghe et al., 2021) and lower than those reported by Perron et al. (2018). The lowest R^2 cv values are due to the weak spatial dependence of the soil properties (Kravchenko, 2003). Furthermore, kriged spatial maps of soil P indices also revealed soil tillage impacts on the sustainable agri-environmental management of P, particularly for identifying zones with agronomic constraints on P fertility, which are associated with the potential loss of P to the environment. For instance, low P fertility values (1% to 4.5%, Figure 3-3b) were observed in the CT field. High P fertility values (5% to 12.5%) were observed in the NT field (Figure 3-4b). At the environmental scale, several NT zones had high P values (8% to 12.5%) that exceeded the national threshold of P saturation (8%) for fine-textured soils (Pellerin et al., 2006; MDDELCC, 2017). Consequently, using the NT system may increase the P fertility level and the risk of P pollution. In addition, visual associations from spatial maps of soil P indices and the other elements (TC, Ca_{M3} and Fe_{M3}) confirmed the influence of these different chemical properties on soil P under the two contrasting tillage practices. Recently, Nze Memiaghe et al (2021) reported similar observations for two contrasting grassland systems in Eastern Canada.

As shown in Table 3-3, the NT system may cause P accumulation in soil and induce significant changes in soil chemical properties. These significant relationships between soil P indices and other chemical properties depend on tillage effects within fields, owing to a decrease in soil pH. In this study, mean pH values from the CT and NT fields were 7 and 6.8, respectively. Piegholdt et al (2013) found similar mean pH values for CT and NT plot soil layers (0-5 cm and 5-25 cm) on Luvisols and Phaeozems in a study on long-term (7 yr) tillage effects. A decrease in soil pH_{H2O} is commonly observed in intensive NT soils compared to CT soils, as reported by Malvezi et al. (2019). This is probably due to the acidifying effects of N fertilizers associated with the decomposition of crop residues (Tarkalson et al, 2006; Limousin and Tessier, 2007; Houx III et al., 2011; Soane et al., 2012; Reeves and Liebig; 2016).

Furthermore, significant relationships were observed between soil pH_{H2O} and both soil P indices (r = 0.23 for P_{M3}, p < 0.01; r = 0.26 for $(P/Al)_{M3}$, p < 0.01), particularly in the 0-5 cm NT soil layer. Similar results were previously reported in intensively cultivated gleysols from this same region (Nolin et al., 1999) and other soils under intensive NT system (Dalchiavon et al., 2017; De Santiago et al., 2019). Other relationships between soil P indices with soil TC, Ca_{M3} and Fe_{M3} were observed under both NT soil layers. This was reported by previous studies (Nolin et al., 1999; Piegholdt et al., 2013; Dalchiavon et al., 2017; De Santiago et al., 2019). Nevertheless, this soil P accumulation in relationship with these soil elements depends also on significant relationships between soil pH_{H2O} and these chemical properties. Thus, significant relationships between soil pH_{H2O} and Ca_{M3} were observed in the 0-5 cm (r= 0.76; p < 0.001) and 5-20 cm (r = 0.67; p < 0.001) NT soil layers. Significant relationships between soil pH_{H2O} and Fe_{M3} in these same layers were (r = -0.68; p < 0.001) and (r = -0.73; p < 0.001), respectively. However, no significant relationships between soil pH_{H2O} and TC were observed in the 0-5 cm (r = 0.07; ns) and 5-20 cm (r = -0.1; ns) NT soil layers.

Finally, the results from Table 3-4 reveal that soil tillage practices induced significant changes in TC, Al_{M3}, Fe_{M3} and Ca_{M3}, affecting the variability of soil P indices. Thus, it is necessary to consider the variability of these chemical properties and their respective spatial structures in order to understand soil P indices variability within fields and according to tillage effects, because these soil chemical properties impact significantly on soil P accumulation under NT systems.

3.7.6. Phosphorus Recommendations and Environmental Implications

As observed in the kriged spatial maps, the soil P accumulation in the first 5 cm of the NT field highlights the environmental risk of P pollution. Average $(P/Al)_{M3}$ values were 2.7% and 7.9% in the CT and NT fields, respectively, after 20 years. Based on current local recommendations (CRAAQ, 2010), P fertilizer applications are 60 kg and 20 kg P_2O_5 ha^{-1} for maize and soybean, respectively, in the CT field. In comparison, the P fertilizer recommendations are 40 kg and 0 kg P205 ha^{-1} for these respective crops in the NT field. Nevertheless, these Quebec local P fertilizer recommendations were developed for uniform P agronomic applications, regardless of field size, soil P variability or their crop productivity potential. Soil variability and potential yield differences in a given field are responsible for large field-scale heterogeneity in the distribution pattern of soil nutrients, such as P (Nolin et al., 1999; Nze Memiaghe et al., 2021).

Variability of soil P indices in both fields ranged from moderate to very high (32% to 60%), indicating that the existing uniform P fertilizer recommendations for maize and soybean are not suited to large (10 ha) fields with contrasting tillage. P fertilization based on uniform recommendations may result both in over and under-fertilization in a large field area (Cambouris et al., 1999). Consequently, accurate P fertilizer recommendations were developed for these fields planted with maize and soybean crops, taking into account

(1) spatial distribution values for soil P resulting from high soil P variability in these large fields and (2) the corresponding Quebec agronomic P recommendations for these respective crops (Figure 3-5).

For instance, accurate P fertilizer recommendations were developed for maize (770 kg P_2O_5) and soybean (460 kg P_2O_5) in the CT field (Table 3-5). In comparison with local P uniform recommendations (648 kg and 216 kg P2O5 for maize and soybean, respectively), this represents an increase in P2O5 of up to 19% and 113% for maize and soybean, respectively. The P fertilizer recommendation for the NT field consisted of a lower amount for maize (366 kg P2O5 ha) and no application (0 kg P_2O_5) for soybean (Table 3-5). This represents a reduction of 14 kg P2O5 for maize and none (0 kg P_2O_5) for soybean, compared to the local uniform P recommendations (380 kg P2O5 and 0 kg P_2O_5). As this example shows, a high CV of soil P in both fields underscores the importance of applying P fertilizer at the right rate and in the right location to avoid creating over- and under-fertilized areas. Thus, there is a potential for controlling spatial variability of soil P in order to increase farm field productivity and yield while reducing P environmental losses.

3.8. Conclusions

This study aimed to examine the impacts of tillage (conventional and no-tillage) on the spatial variability of soil-available P at field scale in a maize-soybean rotation in Eastern Canada with the ultimate goal of improving P fertilizer recommendations. The results revealed that the soil P indices were similar in both layers (0-5 cm and 5-20 cm) under CT. However, in the NT field, these P parameters were higher in the 0-5 cm layer relative to the 5-20 cm layer. These findings have implications for the sustainable management of soil P. The NT system increased soil P $[(P/Al)_{M3}]$ accumulation (7.9%) compared to the CT method (2.7%), presenting a greater risk of P pollution, as observed in the kriged spatial maps. Relationships between soil-available P indices and other chemical properties differed between the contrasting tillage practices.

The variability of soil P indices in both fields ranged from moderate to very high (32% to 60%), indicating that the uniform P fertilizer recommendations currently used for maize-soybean rotations are not suitable for large (10 ha) fields. The use of kriged maps demonstrated the importance of applying P at the right rate in the right location and using this information to develop accurate future P recommendations. The geostatistical models traditionally used to derive soil P indices are not suited to NT fields. Further research should be conducted using other, improved models for better prediction of the spatial structure of soil P indices in NT fields.

3.9. Acknowledgements

Funding for this study was provided by Agriculture and Agri-Food Canada. We thank Claude Lévesque, Sylvie Michaud, and Marc Duchemin for their field assistance and technical support.

3.10. References

Abdi, D., Cade-Menun, B.J., Ziadi, N., and Parent, L.E. (2014). Long-term impact of tillage practices and phosphorus fertilization on soil phosphorus forms as determined by 31 P nuclear magnetic resonance spectroscopy. *Journal of Environmental Quality* 43, 1431-1441.

Allaire, S. E.; van Bochove, E.; Denault, J.-T.; Dadfar, H.; Thériault, G.; Charles, A.; De Jong, R. (2011). Preferential pathways of phosphorus movement from agricultural land to water bodies in the Canadian Great Lakes basin: A predictive tool. *Canadian Journal of Soil Science* 91, (3), 361-374.

Alvear, M.; Rosas, A.; Rouanet, J.; Borie, F. (2005). Effects of three soil tillage systems on some biological activities in an Ultisol from southern Chile. *Soil Tillage Research* 82, (2), 195-202.

Andraski, T. W.; Bundy, L. G.; Kilian, K. C. (2003). Manure history and long-term tillage effects on soil properties and phosphorus losses in runoff. *Journal of Environmental Quality* 32, (5), 1782-1789.

Bertol, I.; Engel, F.; Mafra, A.; Bertol, O.; Ritter, S. (2007). Phosphorus, potassium and organic carbon concentrations in runoff water and sediments under different soil tillage systems during soybean growth. *Soil Tillage Research* 94, (1), 142-150.

Beven, K. J.; Kirkby, M. J. (1979). A physically based, variable contributing area model of basin hydrology. *Hydrological Sciences Journal* 24, (1), 43-69.

Blevins, R.; Frye, W. (1993). Conservation tillage: an ecological approach to soil management. *Advances in Agronomy* 51, 33-78.

Borges, R.; Mallarino, A. (1997). Field-scale variability of phosphorus and potassium uptake by no-till corn and soybean. *Soil Science Society of America Journal* 61, (3), 846-853.

Borges, R.; Mallarino, A. (1998). Significance of spatially variable soil phosphorus and potassium for early growth and nutrient content of no-till corn and soybean. *Communication in Soil Science and Plant Analysis* 29, 2589-2605.

Cade-Menun, B. J.; Carter, M. R.; James, D. C.; Liu, C. W. (2010). Phosphorus forms and chemistry in the soil profile under long-term conservation tillage: A phosphorus-31 nuclear magnetic resonance study. *Journal of Environmental Quality* 39, (5), 1647-1656.

Cambardella, C.; Moorman, T.; Parkin, T.; Karlen, D.; Novak, J.; Turco, R.; Konopka, A. (1994). Field-scale variability of soil properties in central Iowa soils. *Soil Science Society of America Journal 58,* (5), 15011511.

Cambouris, A.; Messiga, A.; Ziadi, N.; Perron, I.; Morel, C. (2017). Decimetric-Scale Two-Dimensional Distribution of Soil Phosphorus after 20 Years of Tillage Management and Maintenance Phosphorus Fertilization. *Soil Science Society of America Journal* 81, (6), 1606-1614.

Cambouris, A. N.; Nolin, M. C.; Simard, R. R. (1999). In *Precision management of fertilizer phosphorus and potassium for potato in Quebec, Canada*, Proceedings of the Fourth International Conference on Precision Agriculture, 1999; Wiley Online Library; pp 847-857.

CRAAQ. (2010). *Fertilization Reference Guide. 2nd edition.* Centre de Référence en Agriculture et Agroalimentaire du Québec: Quebec, QC, Canada, p. 473.

Dalchiavon, F. C.; Rodrigues, A. R.; de Lima, E.; Lovera, L. H.; Montanari, R. (2017). Spatial variability of chemical attributes of soil cropped with soybean under no-tillage. *Revista de Ciências Agroveterinárias* 16, (2), 144-154.

Daverede, I.; Kravchenko, A. Hoeft, R.; Nafziger, E. D.; Bullock, D.; Warren, J.; Gonzini, L. (2003). Phosphorus runoff: Effect of Tillage and Soil Phosphorus Levels. *Journal of Environmental Quality* 32, (4), 14361444.

De Santiago, A.; Recena, R.; Perea-Torres, F.; Moreno, M. T.; Carmona, E.; Delgado, A. (2019). Relationship of soil fertility to biochemical properties under agricultural practices aimed at controlling land degradation. *Land Degradation Development* 30, (9), 1121-1129.

Dick, R. (1992). A review: long-term effects of agricultural systems on soil biochemical and microbial parameters. *Agriculture, Ecosystems and Environment* 40, (1-4), 25-36.

Duiker, S. W.; Beegle, D. B. (2006). Soil fertility distributions in long-term no-till, chisel/disk and moldboard plow/disk systems. *Soil Tillage Research* 88, (1-2), 30-41.

Fernández, F. G.; Schaefer, D. (2012). Assessment of soil phosphorus and potassium following real time kinematic-guided broadcast and deep-band placement in strip-till and no-till. *Soil Science Society of America Journal* 76, (3), 1090-1099.

Friedrich, T.; Derpsch, R.; Kassam, A. (2012). Overview of the global spread of conservation agriculture. In *Field Actions Science Reports*, Special Issue 6.

Guan, D.; Al-Kaisi, M. M.; Zhang, Y.; Duan, L.; Tan, W.; Zhang, M.; Li, Z. (2014). Tillage practices affect biomass and grain yield through regulating root growth, root-bleeding sap and nutrients uptake in summer maize. *Field Crops Research* 157, 89-97.

Heathwaite, A. L.; Quinn, P.; Hewett, C. J. (2005). Modelling and managing critical source areas of diffuse pollution from agricultural land using flow connectivity simulation. *Journal of Hydrology* 304, (1-4), 446461.

Hendershot, W. H.; Lalande, H.; Duquette, M. (2008). Soil reaction and exchangeable acidity. In *Soil sampling and methods of analysis*, Taylor & Francis, B. R., FL, Ed. Taylor & Francis, Boca Raton, FL; pp 173178.

Holm, F.; Zentner, R.; Thomas, A.; Sapsford, K.; Légère, A.; Gossen, B.; Olfert, O.; Leeson, J. (2006). Agronomic and economic responses to integrated weed management systems and fungicide in a wheat-canola- barley-pea rotation. *Canadian Journal of Plant Science* 86, (4), 1281-1295.

Houx III, J.; Wiebold, W.; Fritschi, F. (2011). Long-term tillage and crop rotation determines the mineral nutrient distributions of some elements in a Vertic Epiaqualf. *Soil Tillage Research* 112, (1), 27-35.

Kassam, A.; Friedrich, T.; Derpsch, R.; Lahmar, R.; Mrabet, R.; Basch, G.; González-Sánchez, E. J.; Serraj, R. (2012). Conservation agriculture in the dry Mediterranean climate. *Field Crops Research* 132, 7-17.

Kitchen, N.; Westfall, D.; Havlin, J. (1990). Soil sampling under no-till banded phosphorus. *Soil Science Society of America Journal* 54, (6), 1661-1665.

Kravchenko, A. (2003). Influence of spatial structure on accuracy of interpolation methods. *Soil Science Society of America Journal* 67, (5), 1564-1571.

Kravchenko, A.; Bollero, G. A.; Omonode, R.; Bullock, D. (2002). Quantitative mapping of soil drainage classes using topographical data and soil electrical conductivity. *Soil Science Society of America Journal* 66, (1), 235-243.

Lafond, G. P.; Walley, F.; May, W.; Holzapfel, C. (2011). Long term impact of no-till on soil properties and crop productivity on the Canadian prairies. *Soil Tillage Research* 117, 110-123.

Li, H. (2017). Long-term impact of tillage on the biogeochemical cycle of phosphorus: analysis of the L'Acadie trial (Quebec, Canada) and modelling. Joint PhD thesis in Soils and Environment. Université Laval and Université de Bordeaux, Quebec.

Li, N.; Xu, J.; Yin, W.; Chen, Q.; Wang, J.; Shi, Z. (2020). Effect of local watershed landscapes on the nitrogen and phosphorus concentrations in the waterbodies of reservoir bays. *Science of the Total Environment* 716, 137132.

Limousin, G.; Tessier, D. (2007). Effects of no-tillage on chemical gradients and topsoil acidification. *Soil Tillage Research* 92, (1-2), 167-174.

Liu, K.; Wiatrak, P. (2012). Corn production response to tillage and nitrogen application in dry-land environment. *Soil Tillage Research* 124, 138-143.

Mallarino, A. P. (1996). Spatial variability patterns of phosphorus and potassium in no-tilled soils for two sampling scales. *Soil Science Society of America Journal* 60, (5), 1473-1481.

Malvezi, K. E. D.; Júnior, L. A. Z.; Guimarães, E. C.; Vieira, S. R.; Pereira, N. (2019). Soil chemical attributes variability under tillage and no-tillage in a long-term experiment in southern Brazil. *Bioscience Journal* 35, (2), 467-476.

MAPAQ. (2020). *Portrait-Diagnostic sectoriel de l'industrie des grains au Québec*. Ministère de ¡'Agriculture de l'Alimentation et des Pêcheries du Québec: Québec, QC, Canada, p. 45.

McBratney, A.; Webster, R. (1983). Optimal interpolation and isarithmic mapping of soil properties: V. Co regionalization and multiple sampling strategy. *Journal of Soil Science* 34, (1), 137-162.

MDDELCC. (2017). *Guide de référence du Règlement sur les exploitations agricoles*. Ministère du Développement durable, de I¹ Environnement et de la Lutte contre les changements climatiques: Québec, QC, Canada, p. 185.

Messiga, A. J.; Ziadi, N.; Morel, C.; Grant, C.; Tremblay, G.; Lamarre, G.; Parent, L.-E. (2012). Long term impact of tillage practices and biennial P and N fertilization on maize and soybean yields and soil P status. *Field Crops Research* 133, 10-22.

Messiga, A. J.; Ziadi, N.; Morel, C.; Parent, L.-E. (2010). Soil phosphorus availability in no-till versus conventional tillage following freezing and thawing cycles. *Canadian Journal of Soil Science* 90, (3), 419-428.

Metwally, M. S.; Shaddad, S. M.; Liu, M.; Yao, R.-J.; Abdo, A. I.; Li, P.; Jiao, J.; Chen, X. (2019). Soil properties spatial variability and delineation of site-specific management zones based on soil fertility using fuzzy clustering in a hilly field in Jianyang, Sichuan, China. *Sustainability* 11, (24), 7084.

Nolin, M.; Caillier, M. (1992). Soil variability. II-Quantification et amplitude. *Agrosol5*, (1), 21-32.

Nolin, M. C.; Lamontagne, L. (1990). *Étude pédologique du comté de Richelieu (Québec)*. Agriculture Canada, Research Branch, p. 287.

Nolin, M. C.; Simard, R.; Cambouris, A.; Beauchemin, S. (1999). In *Specific Variability of Phosphorus Status and Sorption Characteristics in Clay Soils of the St-Lawrence Lowlands (Quebec)*, Proceedings of the Fourth International Conference on Precision Agriculture, Wiley Online Library: 1999; pp 395-406.

Nze Memiaghe, J. D.; Cambouris, A. N.; Ziadi, N.; Karam, A.; Perron, I. (2021). Spatial variability of soil phosphorus indices under two contrasting grassland fields in Eastern Canada. *Agronomy* 11, (1), 24.

Olson, K. R.; Ebelhar, S. A. (2009). Impacts of conservation tillage systems on long-term crop yields. *Journal of Agronomy* 8, 14-20.

Pellerin, A.; Parent, L.-É.; Fortin, J.; Tremblay, C.; Khiari, L.; Giroux, M. (2006). Environmental Mehlich-III soil phosphorus saturation indices for Quebec acid to near neutral mineral soils varying in texture and genesis. *Canadian Journal of Soil Science* 86, (4), 711-723.

Perron, I.; Cambouris, A. N.; Chokmani, K.; Vargas Gutierrez, M. F.; Zebarth, B. J.; Moreau, G.; Biswas, A.; Adamchuk, V. (2018). Delineating soil management zones using a proximal soil sensing system in two commercial potato fields in New Brunswick, Canada. *Canadian Journal of Soil Science* 98, (4), 724737.

Piegholdt, C.; Geisseler, D.; Koch, H. J.; Ludwig, B. (2013). Long-term tillage effects on the distribution of phosphorus fractions of loess soils in Germany. *Journal of Plant Nutrition and Soil Science* 176, (2), 217-226.

Pittelkow, C. M.; Liang, X.; Linquist, B. A.; Van Groenigen, K. J.; Lee, J.; Lundy, M. E.; Van Gestel, N.; Six, J.; Venterea, R. T.; Van Kessel, C. (2015). Productivity limits and potentials of the principles of conservation agriculture. *Nature Climate Change* 517, (7534), 365-368.

Pohlmann, H. (1993). Geostatistical modelling of environmental data. *Catena* 20, (1-2), 191-198.

Puustinen, M.; Koskiaho, J.; Peltonen, K. (2005). Influence of cultivation methods on suspended solids and phosphorus concentrations in surface runoff on clayey sloped fields in boreal climate. *Agriculture, Ecosystems and Environment* 105, (4), 565-579.

Reeves, J. L.; Liebig, M. A. (2016). Depth matters: soil pH and dilution effects in the northern Great Plains. *Soil Science Society of America Journal* 80, (5), 1424-1427.

Roberts, W. M.; Gonzalez-Jimenez, J. L.; Doody, D. G.; Jordan, P.; Daly, K. J. (2017). Assessing the risk of phosphorus transfer to high ecological status rivers: Integration of nutrient management with soil geochemical and hydrological conditions. *Science of the Total Environment* 589, 25-35.

Robertson, G. (2008). *GS: Geostatistics for the Environmental Sciences*. Gamma Design Software, Plainwell, MI.

Rodrigues, M.; Pavinato, P. S.; Withers, P. J. A.; Teles, A. P. B.; Herrera, W. F. B. (2016). Legacy phosphorus and no tillage agriculture in tropical oxisols of the Brazilian savanna. *Science of the Total Environment* 542, 1050-1061.

Salekin, S.; Bloomberg, M.; Morgenroth, J.; Mason, D. F.; Mason, E. G. (2021). Within-site drivers for soil nutrient variability in plantation forests: A case study from dry sub-humid New Zealand. *Catena* 200, 105149.

SAS Institute. (2010). *SAS user's guide. Statistics. Version 9.3;* SAS Institute: Cary, NC, USA.

Selles, F.; McConkey, B.; Campbell, C. (1999). Distribution and forms of P under cultivator-and zero-tillage for continuous-and fallow-wheat cropping systems in the semi-arid Canadian prairies. *Soil Tillage Research* 51, (1-2), 47-59.

Soane, B. D.; Ball, B. C.; Arvidsson, J.; Basch, G.; Moreno, F.; Roger-Estrade, J. (2012). No-till in northern, western and south-western Europe: A review of problems and opportunities for crop production and the environment. *Soil Tillage Research* 118, 66-87.

Sörensen, R.; Zinko, U.; Seibert, J. (2006). On the calculation of the topographic wetness index: evaluation of different methods based on field observations. *Hydrology and Earth System Sciences* 10, (1), 101112.

Statistics Canada (2016). *Farm and farm operator data*; Statistics Canada: Ottawa, ON, Canada.

Statistics Canada (2016). *Table 32-10-0408-01: Tillage and seeding practices, Census of Agriculture, 2011 and 2016*; Statistics Canada: Ottawa, ON, Canada.

Sun, W.-X.; Huang, B.; Qu, M.-K.; Tian, K.; Yao, L.-P.; Fu, M.-M.; Yin, L.-P. (2015). Effect of Farming Practices on the Variability of Phosphorus Status in Intensively Managed Soils. *Pedosphere 25,* (3), 438-449.

Tarkalson, D. D.; Hergert, G. W.; Cassman, K. G. (2006). Long-term effects of tillage on soil chemical properties and grain yields of a dryland winter wheat-sorghum/corn-fallow rotation in the Great Plains. *Agronomy Journal* 8:26-33.

Tunney, H. (1990). A note on a balance sheet approach to estimating the phosphorus fertiliser needs of agriculture. *Irish Journal of Agricultural Research:* 149-154.

Tyler, D.; Howard, D. (1991). Soil sampling patterns for assessing no-tillage fertilization techniques. *Journal of fertilizer issues* 8: 52-56.

Vasu, D.; Singh, S.; Sahu, N.; Tiwary, P.; Chandran, P.; Duraisami, V.; Ramamurthy, V.; Lalitha, M.; Kalaiselvi, B. (2017). Assessment of spatial variability of soil properties using geospatial techniques for farm level nutrient management. *Soil Tillage Research* 169, 25-34.

Vieira, S.; Hatfield, J.; Nielsen, D.; Biggar, J. (1983). Geostatistical theory and application to variability of some agronomical properties. *Hilgardia* 51, (3), 1-75.

Wang, C.; Nolin, M.; Wu, J. (1995). In *Microrelief and spatial variability of some selected soil properties on an agricultural benchmark site in Quebec, Canada,* Site-specific management for agricultural systems, Wiley Online Library; pp 339-350.

Webster, R.; Oliver, M. A. (1990). *Statistical methods in soil and land resource survey.* Oxford University Press (OUP).

Whelan, B.; McBratney, A. (2000). The "null hypothesis" of precision agriculture management. *Precision Agriculture* 2, (3), 265-279.

Ziadi, N.; Tran, T. (2008). Mehlich 3-Extractable Elements. In *Soil Sampling and Methods of Analysis,* R. Carter E. G. Gregorich ed; Taylor & Francis: Boca Raton, FL, pp 81-88.

3.11. Tables and Figures

Table 3-1. Mean air temperature and precipitation during the growing season (May-November) in 2014 and 30yr normal (1981-2010) based on the Ste Madeleine weather station (45°37' N, 73°08' W), Montérégie region.

	Precipitation(mm)		Air Temperature (^{o} C)	
	2014	**30-yr normal**	**2014**	**30-yr normal**
May	95.8	85.6	14.2	13.4
June	176.4	97.1	19.3	18.6
July	69.8	102.5	21.1	20.6
August	61.8	98.7	19.3	19.5
September	45.2	87.2	14.7	15.1
October	119.6	103.8	10.1	8.2
November	59.3	101.2	5.3	4.0
Total	627.9	676.1	-	-
Average	-	-	14.9	13.9

Table 3-2. Geostatistical parameters of the soil chemical properties for two soil layers (0-5 cm and 5-20 cm) from the CT and NT fields.

0-5 cm					**5-20 cm**			
	Model[1]	**Sill ratio**[2] (%)	**Range**[3] **(m)**	**R**^{2} **cv**[4]	**Model**	**Sill ratio** (%)	**Range (m)**	**R cv**^{2}
Conventional tillage								
pHH2O	Sph	33	58	0.11	Linen	28	183	0.18

[1] Semivariogram model: Exp: exponential, Lin: Linear, PN: pure nugget, Sph: spherical
[2] Sill ratio (%) = [C/(co+c)] x100; this ratio measures spatial dependence or structure according to Whelan and McBratney (2000)
[3] Distance at which a semivariance becomes constant
[4] Coefficient of determination of cross-validation

TC	PN	-	-	-	PN	-	-	-
P$_{M3}$	Sph	36	50	0.15	Sph	33	51	0.17
Al$_{M3}$	Sph	85	180	0.69	Sph	52	55	0.33
Fe$_{M3}$	Sph	39	60	0.12	Sph	29	48	0.01
Ca$_{M3}$	PN	-	-	-	PN	-	-	-
(P/Al)$_{M3}$	Sph	23	70	0.15	Sph	31	46	0.15
No-tools								
pHH2O	Sph	86	57	0.29	Sph	99	57	0.30
TC	Sph	45	61	0.07	Exp	92	86	0.09
P$_{M3}$	Sph	43	55	0.06	Sph	33	77	0.16
Al$_{M3}$	Sph	52	64	0.27	Sph	52	77	0.24
Fe$_{M3}$	Sph	71	69	0.30	Sph	99	55	0.36
Ca$_{M3}$	Sph	99	77	0.26	Sph	45	90	0.17
(P/Al)$_{M3}$	Sph	46	60	0.08	Sph	45	66	0.17

Table 3-3. Spearman correlation coefficients of the soil-available phosphorus indices (P$_{M3}$) and (P/Al)$_{M3}$ in relation to soil chemical and topography properties for two soil layers (0-5 cm and 5-20 cm) in the CT and NT fields.

0-5 cm

5-20 cm (P/Al)$_{M3}$

	P$_{M3}$		(P/Al)$_{M3}$		P$_{M3}$			
Conventional tillage								
Soil pHH2o	0.06	ns	0.11	ns	0.11	ns	0.13	ns
TC	0.65	***	0.62	***	0.71	***	0.69	***
Al$_{M3}$	-0.09	ns	na		-0.12	ns	na	
Fe$_{M3}$	0.52	***	0.43	***	0.53	***	0.49	***
Ca$_{M3}$	0.41	***	0.40	***	0.28	***	0.28	***
Elevation	0.09	ns	0.1	ns	na		na	
Topographic wetness index	-0.03	ns	-0.1	ns	na		na	
No-tools								
Soil pHH2O	0.23	**	0.26	**	0.00	ns	0.00	ns
TC	0.35	***	0.32	***	0.36	***	0.33	***
Al$_{M3}$	-0.04	ns	na		-0.03	ns	na	
Fe$_{M3}$	0.15	ns	0.13	ns	0.33	***	0.33	***
Ca$_{M3}$	0.26	**	0.30	***	0.13	ns	0.14	ns
Elevation	-0.07	ns	-0.12	ns	na		na	
Topographic wetness index	-0.13	ns	-0.16	ns	na		na	

Significance of correlation indicated by *, **, ***, and ns are equivalent to *p-value* < 0.05, $p < 0.01$, $p < 0.001$, and non-significant respectively; na: non-available

Table 3-4. Multiple regression equations calculated for soil P indices for two soil layers (0-5 cm and 5-20 cm) in the CT and NT fields.

Regression equations	Soil chemical properties selected by the spatial regression				R^2
Conventional tillage	TC^1	$FeM3^2$	$A1M3^3$	$CaM3^4$	
PM3(0-5 cm) =	1.62 +0.805 TC $p < 0.0001$	+0.01013 FeM3 $p < 0.0001$	-0.0035 A1M3 $p < 0.0001$	+0.0005 CaM3 $p < 0.0001$	0.720***
PM3(5-20 cm)=	1.62 +1,378 TC $p < 0.0001$	+0.0085 FeM3 $p < 0.0001$	-0.0032 AlM3 $p < 0.0001$	+0.0003 CaM3 $p < 0.0001$	0.678***
(P/AI)M3(0-5 cm)=	-3.46 +1,042 TC $p < 0.0001$	+0.0053 FeM3 $p < 0.0001$		+0.0005 CaM3 $p < 0.0001$	0.453***
(P/AI)M3(5-20 cm)=	-3.46 +1,408 TC $p < 0.0001$	+0.0051 FeM3 $p < 0.0001$		+0.0002 CaM3 $p < 0.0001$	0.526***
No-tools					
PM3(0-5 cm) =	1.84	+0.005 FeM3	+0.0004 CaM3 $p < 0.0001$	+0.0004 CaM3	0.285***
PM3(5-20 cm)=	1.84	$p < 0.0001$ +0.004 FeM3	$p < 0.0001$ +0.0041 FeM3	$p < 0.0001$ +0.0004 CaM3	0.375***
(P/AI)M3(0-5 cm)=	-0.28	$p < 0.0001$ +0.0031 FeM3		$p < 0.0001$ +0.0004 CaM3	0.319***
(P/AI)M3(5-20 cm)=	-0.28	$p < 0.0001$		$p < 0.0001$	0.392***

Significance of regression indicated by *, **, and *** is equivalent to *p-value* < 0.05, $p < 0.01$, and $p < 0.001$ respectively
[1]TC: Total carbon was analyzed with an Elementar Vario MAX CN analyzer
[2]FeM3: Iron extracted using Mehlich-3 solution
[3]AlM3: Aluminum extracted using Mehlich-3 solution
[4]CaM3: Calcium extracted using Mehlich-3 solution

Table 3-5. Accurate P fertilizer recommendations (kg P2O5) calculated for a maize-soybean rotation based on the kriged value (P/A1)M3 of each specific area (A) compared to the mean value (P/A1)M3 from two soil layers (0-5 cm and 5-20 cm) in the CT and NT fields.

Maize	Ai^5 (ha)	$Ratei^6$ (kg P3 Os ha^)[1]	Qi^{789} (AixRi)	A_3 (ha)	Rate2 (kg P3 Os ha^)[1]	Q_2 (A2χR2)	A_3 (ha)	$Rate_3$ (kg P3 Osha')[1]	Q_3 (A3 χR)3	Total P3 Os[4] Q1+Q2+Q3 (kg P2O)5	P3 Os[5] based on (P/A1)M3 mean value (kg P2O)5
Conventional tillage											
(0-5 cm)	6.1	80	488	4.7	60	282	-	-	-	770	648
(5-20 cm)	8.8	80	704	2	60	120	-	-	-	824	864
No-tools											
(0-5 cm)	8.8	40	352	0.7	20	14				366	380
(5-20 cm)	2.6	60	156	6.9	40	276				432	380

Soya	Ai L⅛L	Ratei (kg P3 Os ha')[1]	Q1 (AixRi)	A_3 (ha)	$Rate_3$ (kg P3 Os ha')[1]	Q_2 (A xR)33	A_3 (ha)	Rate Q33 (kg P3 Osha')[1]	(A xR)33	Total P3 Bones Qt+Q +Q33 (kg P3 Os)	P3 Bone based on (P/A1)M3 mean value (kg P3 Bone)
Conventional tillage											
(0-5 cm)	6.1	60	366	4.7	20	94	460				216
(5-20 cm)	8.8	60	528	2	20	40	568				648
No-tools											
(0-5 cm)	3	0	0	6.5	0	0		--		-0	0
(5-20 cm)	2.6	20	52	6	0	0	0.9	0		052	0

[5] A: specific area (ha) from each specific management zone measured using ArcGIS
[6] Rate: P specific recommendation (kg P2O5 ha^1) corresponding to each kriged value of (P/AI)M3 from the specific area
[7] Q=A X R: this value represents the accurate total P recommendation (kg P2O5) applied in each delimited area (A), based on each kriged value of (P/AI)M3
[8]Total P fertilizer recommendation (kg P2O5) or sum of (Q1+Q2+Q3) applied from all specific areas Ai, A2 and A3
[9] Local P fertilizer recommendation (kg P2O5) from the mean value of (P/AI)M3 traditionally applied in the field according to the Guide de Référence en Fertilisation du Québec (CRAAQ, 2010)

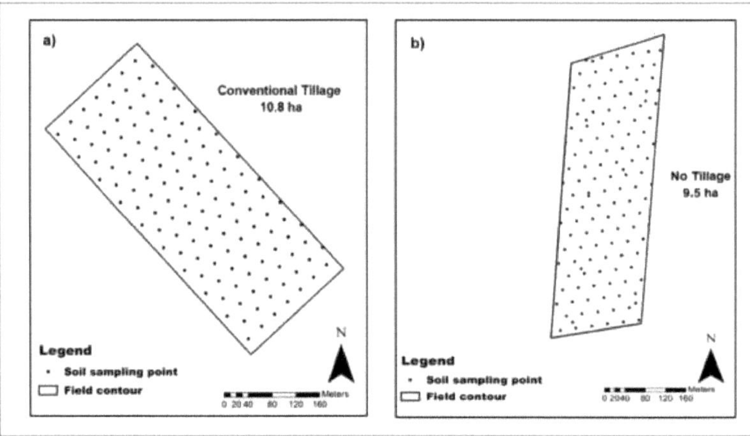

Figure 3-1. Location and sampling strategies of (a) the conventional tillage and (b) no-tillage fields.

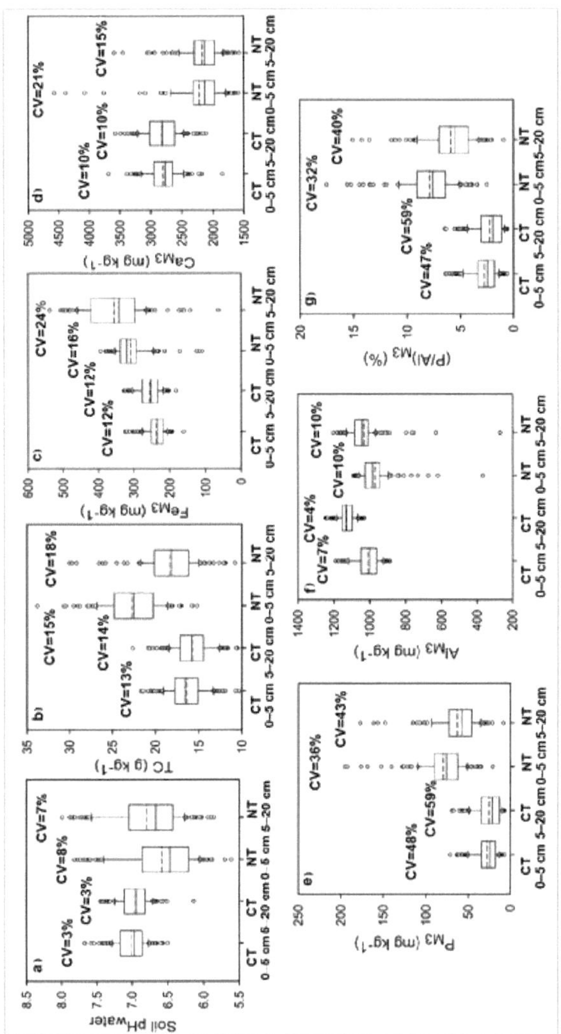

Figure 3-2. Descriptive statistics of soil pHH2O (a), TC (b), FeM3 (c), CaM3 (d), PM3 (e), AlM3 (f), and (P/Al)M3 (g) for two soil layers (0-5 cm and 5-20 cm) in the conventional and no-tillage fields.

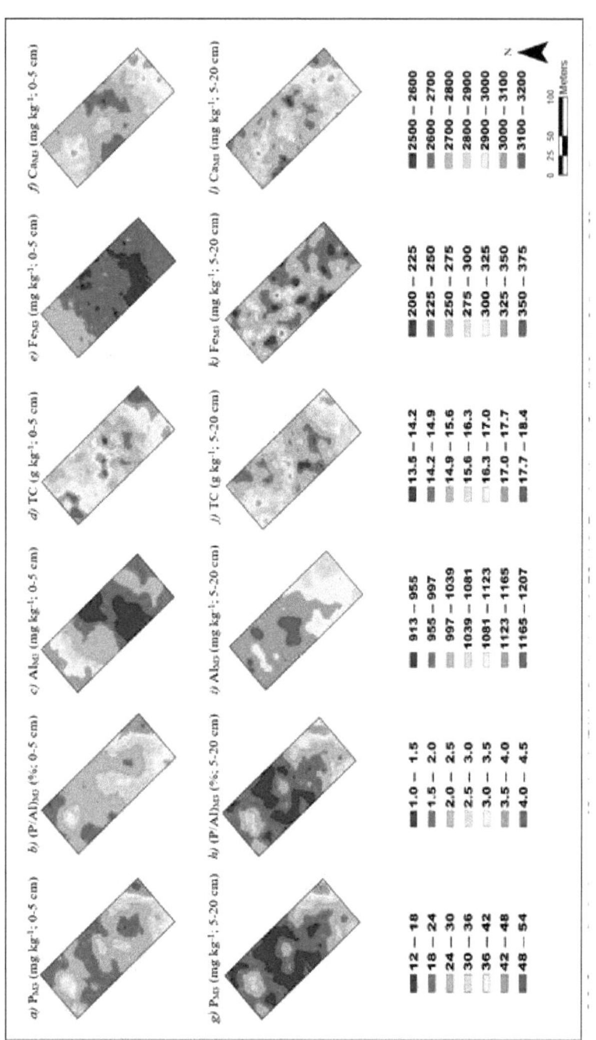

Figure 3-3. Spatial distribution maps of PM3 (a, g), (P/Al)M3 (b, h), AlM3 (c, i), TC (d, j), FeM3 (e, k), and CaM3 (f, l) for the 0-5 cm and 5-20 cm soil layers, respectively, in the conventional tillage field.

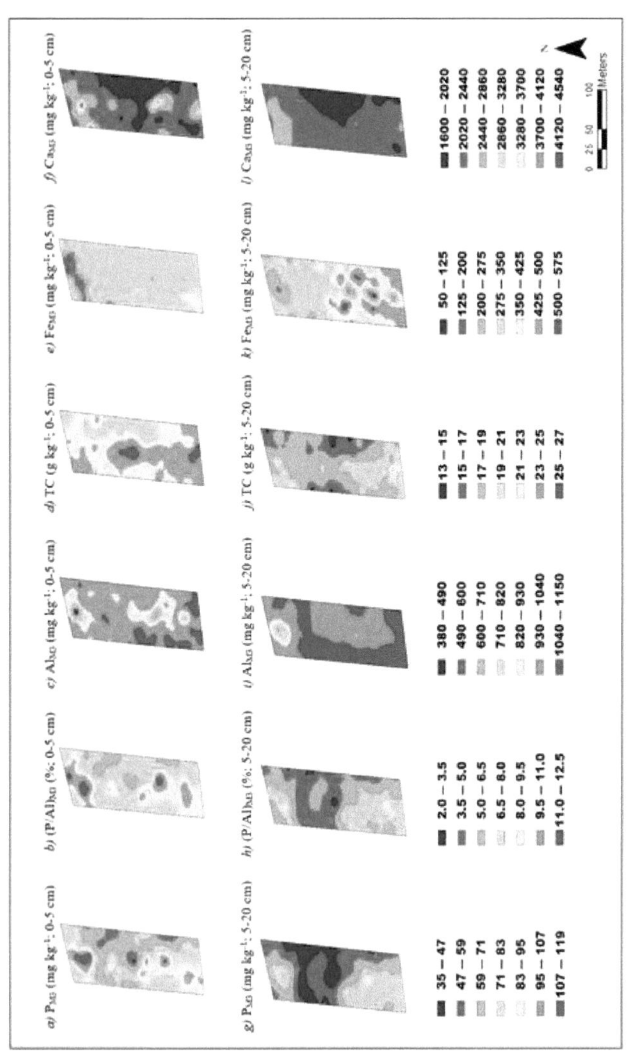

Figure 3-4. Spatial distribution maps of PM3 (a, g), (P/Al)M3 (b, h), AlM3 (c, i), TC (d, j), FeM3 (e, k), and CaM3 (f, l) for the 0-5 cm and 5-20 cm soil layers, respectively, in the no-tillage field.

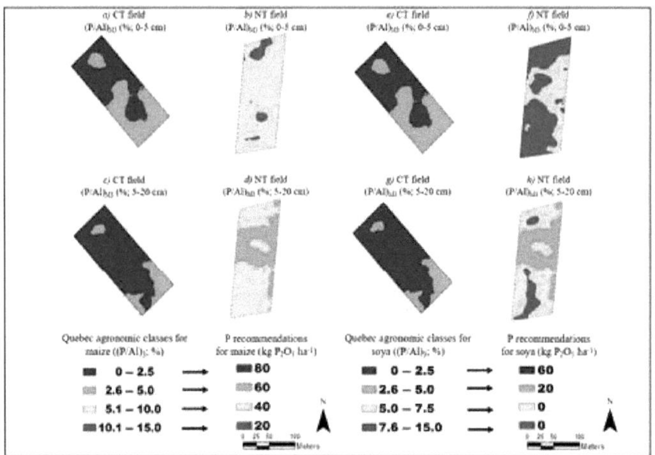

Figure 3-5. Quebec agronomic classes $(P/Al)_{M3}$ and the corresponding accurate P recommendations (kg P2O5 ha⁻ 1) for maize and soya in the conventional and no-tillage fields

Management Zone Delineation Strategies for Soil Phosphorus Recommendations Under a No Till Field in Eastern Canada

J.D. Nze Memiaghe[1,2] , A.N. Cambouris[1]†††, M. Duchemin[1] , N. Ziadi[1] and A. Karam[2]

[1] *Agriculture and Agri-Food Canada, Quebec Research and Development Centre, 2560 Hochelaga Boulevard, Quebec City, QC G1V 2J3, Canada. jeff-nze.memiaghe@agr.gc.ca; athyna.cambouris@.agr.gc.ca; marc.duchemin2@agr.gc.ca; noura.ziadi@agr.gc.ca*

[2]*Soils and Agri-Food Engineering Department, Lavai University, 2425 rue de l'Agriculture, Quebec City, QC G1V 0A6, Canada; jdnzm@ulaval.ca; antoine.karam@fsaa.ulaval.ca*

This paper was submitted to Canadian Journal of Soil Science.

4.1. Core Ideas

•   Despite a low pedodiversity, variability of soil P was moderate (32-36%) under large no tillage field, meaning that P fertilizer recommendations cannot be adapted uniformly.

•   A significant relative correlation $_{ECa30\text{-}soil}$ P was observed under no tillage field, generating three management zones (MZs) delineation strategies.

4.2. Summary

The spatial variability of soil phosphorus (P) under direct seeding (DS) must be considered in order to develop P fertilisation programmes that avoid P losses and stratification. Little information exists on the control of spatial variability of field P using management zones (ZAs), particularly in soils under DS. The objective of this study was to delineate the ZAs of a field under SD (9.5 ha) for 20 years, grown in a corn-soybean rotation, by exploring the soil apparent electrical conductivity (ECa)-P relationship in order to reduce the spatial variability of soil P for site-specific P fertilizer recommendations in eastern Canada. To resolve this, an intensive 35 m by 35 m sampling grid (total: 134 soil samples) was carried out in autumn 2014. Mehlich-3 (M3)-extractable P and aluminium (Al) were determined in the top soil layer (0-5 cm). The (P/Al)$_{M3}$ indicator was then calculated. ECa data were measured at two depths (0-30; 0-100 cm) using the Veris system. ZAs were delineated using (P/A1)$_{M3}$ or CEa alone or in combination. The average (P/Al)$_{M3}$ content was 7.9%. Soil P variability was moderate (32-36%), indicating that uniform P fertilizer recommendations cannot be adapted to this large-area field under SD. Mean values of $_{CEa30}$ and $_{CEa100}$ were 15.8 and 32.6 mS m^{-1} , respectively. A significant soil $_{ECa30\text{-}P}$ correlation (0.22-0.23) was observed. Delineation into two or three ZAs using (P/Al)$_{M3}$ measurements represented the best agronomic strategy, generating the highest P reductions (40-74 kg $_{P2O5}$). This study highlighted the importance of a strategy to reduce the spatial variability of soil P, by delineating ZAs, while reducing P losses from the field.

Keywords: soil electrical conductivity, ISODATA method, Veris, spatial variability, variance reduction, precision agriculture.

4.3. Abstract

Soil phosphorus (P) spatial variability under no-tillage (NT) management should be taken into account for developing P fertilization programs while avoiding P losses and stratification. Little is known on controlling field-scale P spatial variability using management zones (MZs), particularly under no-tilled soils. The general objective of this study was to delineate MZs in a corn-soybean rotation in a cropland (9.5 ha) under NT for 20 years, exploring the soil apparent electrical conductivity (ECa)-P relationship to reduce soil P spatial variability for site-specific P fertilizer recommendations in Eastern Canada. To address this, an intensive grid sampling of 35 m by 35 m (total: 134 soil samples) was implemented in fall 2014. Mehlich-3 extractable P and aluminum (Al) were

---

††† There's a potential for reducing spatial variability of soil P, by delineating MZs, for sustainable P2O5

recommendations under a no tillage field.

determined in the upper soil profile (0-5 cm). The (P/Al)$_{M3}$ index was then calculated. The ECa data was measured in two depths (0-30 cm and 0-100 cm; $ECa30$ and $ECa100$) using a Veris system. The MZs were delineated using (P/Al)$_{M3}$ or the ECa alone or in combination. Mean (P/A1)$_{M3}$ value was 7.9%. Variability of soil P was moderate (32% to 36%), indicating that uniform P fertilizer recommendation is not adapted to this large NT field. Mean values of $ECa30$ and $ECa100$ were 15.8 and 32.6 mS m^{-1} , respectively. A significant $ECa30\text{-}soil$ P correlation (0.22-0.23) was observed. Delineating two to three MZs using (P/Al)$_{M3}$ measurements represented the best agronomic strategy, generating the highest P reductions (40-74 kg $P2O5$). This study highlighted a potential for reducing spatial variability of soil P, by delineating MZs, while reducing P losses from crop fields...

Keywords: soil electrical conductivity, ISODATA method, Veris, spatial variability, variance reduction, precision agriculture.

4.4. Introduction

Corn (*Zea mays* L.) and soybean (*Glycine max* L.) represent the dominant legume-cereal association in North America, corresponding to 75% of cultivated land and 90% of agricultural production in Eastern Canada (MAPAQ 2020). Owing to their high economic values, these crops are cultivated using conservation tillage practices such as no-tillage (NT) to promote sustainable farming systems (Kassam et al. 2012; Soane et al. 2012). Nevertheless, many agri-environmental issues have been ascertained regarding NT agronomic practice in relation to P stratification (Abdi et al. 2014; Messiga et al. 2010), transport and runoff to surface waters (Puustinen et al. 2005) and causing P eutrophication in rivers (Patoine et al. 2017). Another major obstacle is the expensive costs related to traditional and intensive soil sampling associated with the laboratory analyses required for accurate mapping of desired soil properties (Bongiovanni and Lowenberg-DeBoer 1999; Bullock and Bullock 2000; Peralta and Costa 2013) such as soil P (Lawrence et al. 2020) with the goal of implementing site-specific P management, particularly under no-tilled soils. Previous authors (Borges and Mallarino 1998; Mallarino 1996) reported major problems for soil sampling, which were especially aggravated under no-tilled fields, caused by the high spatial variability of soil P. These major concerns are due to the limited mixing of soil, crop residues, and P fertilizers. Thus, we assume that it is important to improve our knowledge of spatial variability of soil P under NT management for developing more rational and cost-effective P fertilization programs that will enhance yields and reduce P losses to the environment. By ignoring spatial variations in soil P fertility, P fertilizer applications could result in P over-fertilization in low-yielding areas and P under-fertilization in high-yielding areas (Schumann et al. 2003).

Delineating management zones (MZs) is the most general approach used to manage within-field spatial variability (Ferguson et al. 2003). The MZ approach delineates smaller homogeneous units in which uniform management can be used and applies variable management among the different units (Zebarth et al. 2012). MZs are sub-regions with similar characteristics affecting yield or with the same yield productivity (Khosla and Shaver 2001; Metwally et al. 2019). The development of MZs divides a big field into subfields having similar properties, where a uniform rate of fertilization can be implemented (Ferguson et al. 2003). Thus, delineating MZs has a great potential to control soil and crop variability (Ameer et al. 2022) by achieving site-specific nutrient management (Li et al. 2008). Delineating MZs represents an excellent alternative for grid soil sampling and is a tool for evaluation of agronomic yields in relation to physico-chemical soil properties for a better crop modeling study (Armstrong et al. 2009; Li et al. 2007; Rab et al. 2009). Thus, use of MZs represents a popular and promising approach in precision agriculture (Ameer et al. 2022; Cambouris et al. 2014). This agronomy strategy may represent a solution aiming to capture or reduce spatial variability of soil P while reducing sampling costs for small farmers.

Moreover, it is necessary to efficiently delineate site-specific MZs (Yao et al. 2014) using auxiliary properties (Cambouris et al. 2006). The auxiliary properties most used in spatial mapping of soil properties are apparent electrical conductivity (Perron et al. 2018), remote sensing data extracted from satellite or airborne images (Sullivan et al. 2005), geophysical measurements such as electrical soil surveying (Adamchuk et al. 2015; Corwin and Lesch 2003), and parameters extracted from model elevation data (Odeh et al. 1995). Nevertheless, these auxiliary properties must present a strong spatial structure as well as a high correlation (r > 0.5) with the main soil property to be mapped (Doerge 1999; Heiniger et al. 2003) in order to be useful for digital mapping of soil properties.

Measuring apparent electrical conductivity (ECa) from electrical resistivity and electromagnetic induction represents the best modern technology for delineating MZs within the field (Ameer et al. 2022). Indeed, this method is characterized by its high measurement density, its precision and its measurement speed. Additionally, ECa can also be measured in a simple and inexpensive way. Thus, ECa is among the most useful and easily obtained spatial properties of soil (Corwin and Lesch 2003). Moreover, ECa can be used as an auxiliary property for better characterization of within-field spatial variability of soil properties (Cambouris et al. 2006). Indeed, ECa maps are important tools for distributing homogeneous areas because of their strong correlation with several soil physico-chemical properties, such as texture (Williams and Hoey 1987), water content (Kitchen et al. 2005), salinity (Heilig et al. 2011), cation exchange capacity (Saifuzzaman et al. 2021), and organic matter content (Becker et al. 2022). Thus, temporally stable ECa data, which are highly related to inherent soil properties, are currently the most effective auxiliary tool for defining and delineating within-field MZ (Fraisse et al. 2001) if specific relationships with the concentrations of desired soil chemical properties can be mapped with minimum error (Omonode and Vym 2006). To this end, Omonode and Vyn (2006) reported relationships between ECa and soil nutrients, such as P, that may help to successfully delineate MZs for a sitespecific P management. Thus, ECa measurements can also provide a useful tool in soil P fertility evaluation for site-specific agriculture because it has the potential to identify areas within fields where soil types, nutrient levels, and productivity differ (Omonode and Vyn 2006). Furthermore, crop yield represents another important auxiliary property in precision agriculture. Yield mapping is a simple, inexpensive tool for monitoring crop yield at fine spatial resolutions (Rodrigues and Corá 2015). Thus, yield mapping can be used by farmers to identify underperforming areas where there may be soil P fertility issues (Higgins et al. 2019).

The $(P/Al)_{M3}$ index represents an important agri-environmental indicator developed in Eastern Canada (Khiari et al. 2000; Pellerin et al. 2006). This index aims to estimate soil P saturation to allow accurate P fertilizer recommendations while integrating agronomical aspects and environmental risks. In Eastern Canada, local P recommendations were developed for uniform P agronomic applications based on the $(P/Al)_{M3}$ index (CRAAQ 2010), regardless of field size, soil P variability, or crop productivity potential. Conversely, Bolinder et al (2000) reported that soil P content represents a major environmental concern. To this end, soil agri-environmental index $(P/Al)_{M3}$ could be managed specifically in each MZ in the Province of Quebec.

Many studies were carried out on the use of ECa measurements for delineating MZs for characterizing the spatial variability of soil properties (Ameer et al. 2022; Peralta and Costa 2013; Perron et al. 2018) such as soil N (Argento et al. 2021; Khosla et al. 2002) in precision agriculture. However to date, there have been no studies conducted on delineating MZs using the ECa-P relationship to characterize spatial variability of soil agri-environmental index $(P/Al)_{M3}$ to reduce soil sampling and spatial variability of soil P for optimizing future P recommendations under NT systems in Eastern Canada. Thus, our hypothesis is that ECa-P relationship can be used to characterize the spatial variability of soil-environmental index $(P/Al)_{M3}$. Measuring ECa would be essential in this field-scale study, because this methodology may improve characterization of spatial patterns of soil agri-environmental index $(P/Al)_{M3}$, which in turn can be used to define MZs. Therefore, delineating $(P/Al)_{M3}$- based homogeneous MZs in relationship with ECa measurements would help to apply more efficiently P fertilizers based on a site-specific soil P management in Eastern Canada.

Thus, this study aimed to delineate MZs using the ECa-P relationship to reduce soil sampling and spatial variability of soil agri-environmental index $(P/Al)_{M3}$ in order to develop accurate P recommendations under NT system in Eastern Canada. The specific objectives were to: (1) investigate spatial variability and structure of soil chemical properties such as P indices (P_{M3}, $[P/Al]_{M3}$) and other selected auxiliary properties ($ECa30$, $ECa100$, elevation, yields) using statistics and geostatistics tools; (2) study spatial relationships between agri- environmental index $(P/Al)_{M3}$ and other selected soil chemical and auxiliary properties, including ECa, based on Pearson correlations and multiple regression equations; (3) determine the optimal number of MZs using the decrease of variance method based on combinations generated from the relationship between ECa and agri- environmental index $(P/Al)_{M3}$ measurements; and (4) validate the MZs in order to optimize P fertilizer recommendations.

4.5. Material and methods

4.5.1. Experimental site

The study was conducted in the Montérégie region (Quebec, Canada). This region is located in the physiographic region of the St. Lawrence Lowlands, which is characterized by very flat landforms (0% to 5%) (Wang et al. 1995). A commercial field under NT practice established in 1994 was located at La Présentation (45°36'N; 73°02'W; 9.5 ha) and referred to as NT field (Fig. 4-1 *a*). The soil texture was a clay loam, classified to the fine-textured group (soil with clay content > 30%). The soil series represented in the field was Kierkoski (Nolin and Lamontagne 1990) and corresponded to an Orthic Humic Gleysol (SISCAN 1998). The field was poorly drained. The mean annual temperature and total annual precipitation were 5.9°C and 984 mm, respectively (Table 4-1).

The field was managed under corn-soybean rotation. Corn was planted with 0.75-m row spacing in early May during the 2012, 2014 and 2016 growing seasons. Soybean was planted with 0.3-m row spacing in early May during the 2013 and 2015 growing seasons. Both crops were harvested in late October. Crop management and fertilization practices were based on local recommendations in the province of Quebec (CRAAQ 2010). The NPK fertilizer sources were as follows: N fertilizer provided from ammonium nitrate (34-00); P fertilizer provided from triple superphosphate (0-46-0); and K fertilizer provided from muriate of potash (00-60). The NPK fertilizers were band applied at seeding. Nitrogen was applied twice by fractionation (at the seeding stage and during the 6-8 leaf stage), while P and K were both applied at the seeding stage following local recommendations. For each crop, application rates were applied based on previous soil chemical analysis and following province of Quebec recommendations (CRAAQ 2010).

4.5.2. Soil sampling and physico-chemical analyses

An intensive soil sampling was performed on November 2014 using a grid design with a sampling interval of 35 m × 35 m (soil sampling density: 14 samples ha^{-1}), providing 134 georeferenced sampling points (Fig. *4-1a*). The georeferenced sampling points and field boundary were marked using a handheld differential global positioning system (DGPS) receiver (Thales Navigation, Santa Clara, CA, USA). A composite soil sample of four cores was taken within a radius of 1 m around each sampling point at the 0-5 cm soil depth using a 5-cm- diameter Dutch auger. This was based on suggestions from previous studies (Cade-Menun et al. 2010; Cambouris et al. 2017) on P stratified soils under NT fields, for sustainable P recommendations. Soil samples were air-dried, ground and sieved through a 2-mm sieve for characterization.

Soil pH (1:1 soil-water ratio) was measured according to Hendershot et al. (2008). Soils were extracted with a soil solution ratio of 1:10 using Mehlich-3 (M3) solution (Ziadi and Tran 2008), and the concentrations (mg kg^{-1}) of P_{M3}, iron (Fe_{M3}), calcium (Ca_{M3}) and aluminium (Al_{M3}) in the extract were determined by inductively coupled plasma optical emission spectroscopy (ICP-OES; Model, 4300DV, PerkinElmer, Shelton, CT, USA). The soil P agri-environmental indicator (P/Al)$_{M3}$ was then calculated. Total carbon (TC, g kg^{-1}) content was measured using an Elementar Vario MAX CN analyzer (Elementar Analysensysteme GmbH, Hanau, Germany).

Soil elevation and the topographic wetness index (TWI; TWI = ln[α/tan β]) were generated using the digital terrain model derived from light detection and ranging (LiDAR) imagery of the province of Quebec. The spatial resolution of these raster images was 1 m. The NT field contour was extracted from the sheet 31H11SE using the "Extract by mask" ArcGIS tool (ArcGIS Software version 10.3; ESRI, Redlands, CA, USA). The value of TWI depends on the upslope area unit contour length (α) and the local slope (tan(β)) (Beven and Kirkby 1979; Sörensen et al. 2006). Mean elevation and TWI values were generated from each sampling point using the buffer method within a 1-m radius and the "Zonal Statistics as Table" ArcGIS tool (ArcGIS Software version 10.3; ESRI, Redlands, CA, USA). Maps generated using LiDAR data aimed to indicate areas at risk of P mobilization and transport in surface runoff pathways at field-scale (Cassidy et al. 2019).

4.5.3. Measurement of soil electrical conductivity data

The ECa data were acquired during the fall of 2014, after the corn harvest, using a commercial Veris 3100 galvanic contact resistivity sensor (Veris Technologies, Inc., Salina, KS, USA) equipped with a Garmin GPS 17x HVS sensor (Garmin International, Olathe, KS, USA) with an accuracy of less than 1 m. This sensor was configured according to the Wenner array using six coulter electrodes (one pair to inject current and two pairs to measure change in electrical potential), as described by Sudduth et al. (2005). Soil ECa data were recorded by the Veris system simultaneously from two depths, 0-30 cm ($ECa30$) and 0-100 cm ($ECa100$) (Kweon et al. 2012; Lund et al. 1999).

The data from the sensor were acquired along parallel transects spaced approximately 10 m apart with 1 Hz logging frequency, corresponding to a measurement every 2-3 m when operating with a speed of approximately 10 km h^{-1}. The data density was about 229 measurements per hectare (Fig. 4-1b). As suggested by Sanches et al. (2018), erroneous ECa data were eliminated by the exclusion of points exceeding three standard deviations.

4.5.4. Measurement of corn and soybean yields

Corn (2012, 2016) and soybeans (2013, 2015) were harvested mechanically on September using combine harvesters. Unfortunately, corn (2014) yield data are missing due to corrupted file. The yields data were measured using on-the-go yield monitors (RDS Technology) installed in the combine harvesters. The system was combined with a DGPS receiver.

Yield measurements were taken using grain sensors, with each measurement covering an area of about 2 m × 6 m (2 m is the average forward distance traveled by a combine during 1 s, and 6 m is the width of the combine header). Simultaneously, the site coordinates (x, y) were determined using a GPS unit (real-time kinematic GPS system; Trimble Navigation, Ltd., Sunnyvale, CA, USA). The moisture content for corn and soybean grain yields was adjusted to 15% and 13%, respectively.

4.5.5. Statistical and geostatistical analyses

Data were analysed using descriptive statistics and geostatistics. Means and coefficient of variation (CV) values were estimated using the PROC UNIVARIATE procedure of SAS software version 9.4 (SAS Institute 2010). The CVs of soil chemical and auxiliary properties were classified based on the approach of Nolin and Caillier (1992) as follows: (1) low (CV < 15%); (2) moderate (15% < CV < 35%); (3) high (35% < CV < 50%); (4) very high (50% < CV <100%); and (5) extremely high (CV > 100%). Descriptive statistics provide the overall variability of soil chemical and auxiliary properties but not the nature of their variability (structured or randomized) or their spatial trend (Ameer et al. 2022; Cambouris et al. 2006).

Geostatistical analyses (parameters calibrations, geostatistical computations, and model validations) were performed to determine spatial structures of soil chemical and auxiliary properties using GS+ version 9 (Robertson 2008). The spatial structure of different properties was evaluated via isotropic and anisotropic semivariograms. Experimental semivariograms, the main component of kriging, are an effective tool for evaluating spatial variability (Wu et al. 2009). Semivariogram parameters for each theoretical model (spherical and exponential) were generated. The corresponding nugget (c_0), partial sill (C), sill (c_{0+C}), and range values of the best-fitting theoretical model were determined. The partial sill ratio [C/(c_{0+C})] was calculated to determine the spatial structure of the soil chemical and auxiliary properties. Semivariograms with a partial sill ratio of < 25%, 25% to 40%, 40% to 60%, 60% to 75%, or >75% were considered to have a low, low moderate, moderate, strong moderate, or strong spatial structure, respectively (Whelan and McBratney 2000). The range estimates the distance beyond which two observations are spatially uncorrelated (Goovaerts 1998).

Spatial variability maps were generated using block kriging interpolation from GS+ version 9 (Robertson 2008) in combination with ArcGIS Software version 10.3 (ESRI, Redlands, CA, USA). The spatial resolution of the raster maps was 1 m^2. These raster maps were circumscribed to the field contours using the "Extract by mask" ArcGIS tool. Spatial maps were produced into five classes similarly to Ameer et al. (2022) for easier comparisons. Kriged map reliability was evaluated using cross-validation analysis (R^2cv) (Kravchenko et al. 2002).

4.5.6. Relationships between soil P agri-environmental index (P/Al)$_{M3}$ and other soil chemical and auxiliary properties

First, the mean values of the auxiliary properties ($ECa30$, $ECa100$, elevation, yields) were generated from interpolated spatial maps and measured within a 5-m radius of the soil sampling locations (buffer method). The buffer method was performed using the "Buffer Geoprocessing" ArcGIS tool. Then, Pearson's correlation coefficients (r) between soil P agri-environmental index (P/Al)$_{M3}$, $ECa30$, $ECa100$ and soil chemical properties ($pHwater$, TC, $PM3$, $AlM3$, $FeM3$, $CaM3$, [P/Al]$_{M3}$) and auxiliary properties (elevation, TWI, yields) were also calculated using the CORR and UNIVARIATE procedures of SAS software version 9.4 (SAS Institute 2010).

To determine spatial relationships between the main response variable (soil P agri-environmental index (P/Al)$_{M3}$

and other explanatory variables (pH_{water}, TC, Fe_{M3}, Ca_{M3}, ECa_{30}, ECa_{100}, elevation, yields), a mixed regression model was used with exponential spatial correlation structure between observations. Since measurements were taken at different spatial points (x_i, y_i) and at the soil depth (0-5 cm) at each point, a different exponential spatial correlation was assumed for this soil depth. Letting c_n and α_0 denote the nugget and the range estimates, respectively, the correlation between any two observations at distance d apart, $p(d)$, is equal to:

$$\rho(d) = (1 - \hat{c}_n) \exp\left(\frac{-d}{\hat{\alpha}_0}\right)$$

The regression models were fitted at soil depth (0-5 cm). For each response variable, the Box-Cox methodology was used to transform the response variable accordingly. Thus, spatial regression equations were generated for $(P/Al)_{M3}$ in relationship with other soil chemical and auxiliary properties, aiming to understand which factors influenced the spatial pattern of $(P/Al)_{M3}$ over time, and evaluate strength and stability of the relationships between $(P/Al)_{M3}$ and these soil chemical and auxiliary properties. A coefficient of determination (R^2) was calculated to evaluate the performance of the model. All analyses were performed using the mixed procedure of SAS software version 9.4 (SAS Institute 2010) at the 0.05 level of significance. The normality assumption was validated using the Shapiro-Wilk's statistic on the normalized residuals of the model.

4.5.7. Delineation of MZs

Delineation of MZs was performed using the ISODATA method. The ISODATA method was applied using the Iso Cluster Unsupervised Classification function of ArcGIS software package version 10.3 (ESRI, Redlands, CA, USA). The ISODATA method is a classification algorithm (Tou et al. 1974) that does not require introduction of class characteristics before classification (unsupervised classifier). This delineating technique calculates the means of classes evenly distributed in space, then aggregates iteratively the remaining data by minimising Euclidean distance of each data to a class average (Guastaferro et al. 2010). Then, each iteration recalculates the means and reclassifies the data relative to the new means. Splitting, merging and deleting iterative classes are based on input parameters, such as minimum and maximum number of classes, maximum number of iterations, and maximum class standard deviation, and then all data are grouped into the nearest class.

Based on relationship values between soil P agri-environmental index $(P/Al)_{M3}$, ECa measurements, and soil chemical and auxiliary properties, delineating MZs was performed using the following three MZ delineation strategies (MZDS): (1) the $(P/Al)_{M3}$; (2) the ECa; and (3) the combination of $(P/Al)_{M3+ECa}$.

4.5.8. Determination and validation of the optimal number of MZs

The inflection points of the variance decrease curves for $(P/Al)_{M3}$, ECa, and other soil chemical and auxiliary properties were estimated to determine the optimal number of MZs for each property using these three strategies, respectively. The optimal number was determined using the variance reduction method (Haghverdi et al. 2015). Using this approach, the total within-zone variance was expressed as a percentage of the variance for the entire field (i.e. a MZ) (Cambouris et al. 2006). Thus, the optimal number of MZs was selected based on the magnitude of reduction in total within-zone variance (Moral et al. 2010; Xin-Zhong et al. 2009).

After the optimal number of MZs was determined, validation of the optimal number of MZs was performed for soil $(P/Al)_{M3}$ and other properties using each strategy. Thus, normality tests of soil $(P/Al)_{M3}$ and other properties were achieved using the PROC UNIVARIATE procedure in SAS (SAS Institute 2010). For normally distributed parameters, an ANOVA combined with a LSD multiple comparison test (p-value < 0.05) was performed to determine if the soil properties varied significantly among the MZs. A mixed regression model with an exponential covariance structure was used. Non-normally distributed parameters were analysed using nonparametric tests (Wilcoxon and Kruskal-Wallis) in the PROC MIXED procedure (SAS Institute 2010).

The significant differences in soil properties within the delineated MZs made it possible to determine how well these three MZDS, such as $(P/Al)_{M3}$, ECa and $(P/Al)_{M3+ECa}$ measurements, captured within-field spatial variability of soil $(P/Al)_{M3}$ and evaluated the delineation efficiency.

Lastly, accurate P fertilizer recommendations were developed in the commercial field for corn and soybean in each MZ using these three strategies. These results were compared to the local uniform P fertilizer recommendations (based on $[P/Al]_{M3}$ mean values) established for Eastern Canada (CRAAQ 2010) to estimate field-scale P_2O_5 profits and losses. The *modus operandi* for evaluating the MZDS using the ISODATA method in order to develop P fertilizer recommendations is summarized in Figure 4-2.

4.6. Results and discussion

4.6.1. Intensity of variation of soil agri-environmental index $(P/Al)_{M3}$ and other soil chemical and auxiliary properties.

In the 0-5 cm layer, CVs of soil chemical properties ranged from low to high (8% to 36%) based on Nolin and Caillier (1992) (Table 4-2). They were classified as follows: low for soil pH_{water} and Al_{M3}; moderate for TC, Fe_{M3} and Ca_{M3}; and moderate to high for soil P indices. P_{M3} and $(P/Al)_{M3}$ values ranged from 21 to 195 mg kg^{-1} and from 2.5% to 18% with means of 80 mg kg^{-1} and 8%, respectively (Table 4-2).

The highest CVs (32% to 36%) were measured for soil P indices (Table 4-2). Previous field-scale studies reported the highest CVs for soil P (Nolin et al. 1996; Peralta and Costa 2013; Quenum et al. 2012) with similar sampling grids (Nolin et al. 1999) under similar crops and tillage practices (Malvezi et al. 2019; Omonode and Vyn 2006). Rodrigues and Corá (2015) reported a mean CV of soil P of 59% in a 12-year NT field under corn-fallow rotation.

In Eastern Canada, high CVs of soil P in agricultural landscapes were generally due to long and narrow plot scales made up of one or more rounded beds (Quenum et al. 2012). Moreover, Perron et al. (2018) reported a high CV of soil P, owing to a greater pedodiversity within the field. Similar results were reported in previous field-scale studies performed on diverse soil series or parent materials (Bogunovic et al. 2014; Han et al. 1996; Miao et al. 2006). However, CVs of soil P indices remain high despite the low pedodiversity (unique soil series) observed in our field-scale study. Consequently, a high intensity of variation of soil P within the same soil series will impact the delineation of P-based homogeneous MZs in a P accumulation region with the goal of guiding intensive soil sampling and specific P fertilizer management (Pierce and Nowak 1999).

The CVs of auxiliary properties ranged from low to moderate (0.3% to 16%) (Table 4-2). They were classified as follows: low for elevation, TWI, ECa_{100} and yield (2012, 2013 and 2016); and moderate for ECa_{30} and yield2015. The highest CVs (16%) were measured for ECa_{30} and yield2015. Elevation and TWI values ranged from 34.2 to 35 m and from 7 to 14 m with means of 34.8 m and 10 m, respectively. This field was flat, as confirmed by previous studies in the region (Wang et al. 1995). As suggested by Wilson et al. (2016), low CVs of elevation and TWI combined with lack of specific significant relationship between soil P and these topographical properties (Fig. 4-5) confirms failure to account for both topographical properties in this study.

The ECa_{30} and ECa_{100} values ranged from 10 to 29 mS m^{-1} and from 20 to 44 mS m^{-1} with means of 15.8 mS m^{-1} and 32.6 mS m^{-1}, respectively (Table 4-2). These values were higher than others (Lajili et al. 2021; Nolin et al. 2003). High ECa values are due to high clay content (Brune et al. 1990; Peralta and Costa 2013), characterizing Gleysolic soil series in the Province of Quebec. Besides, Gleysolic soils series were located in the St-Lawrence physiographic valley with glacio-marine sediments, as parent material. Glacio-marine sediments are characterized by a high salt content. CVs for ECa measurements ranged from low to moderate (9% to 16%). These results were lower than others (Nolin et al. 2003; Peralta and Costa 2013) because of the low pedodiversity within this field. High CVs for ECa measurements resulted from (1) high pedodiversity and (2) high variable conditions of soil surface moisture during time of data acquisition in the field (Omonode and Vyn 2006; Perron et al. 2018).

Corn yield2012 and yield2016 values ranged from 5.2 to 13.9 Mg ha^{-1} and from 9.1 to 18.8 Mg ha^{-1} with means of 11.1 and 12.8 Mg ha^{-1}, respectively (Table 4-2). These mean yields represented a 15% difference during these two growing seasons. CVs for corn yields were low (8% to 9%). Therefore, temporal CV of corn yield was 12.5% during these two growing seasons. Soybean yield2013 and yield2015 values ranged from 2.3 to 4.6 Mg ha^{-1} and from 1.5 to 6.1 Mg ha^{-1} with means of 3.6 and 3.2 Mg ha^{-1}, respectively (Table 4-2). These mean yields represented an 11% difference during these two growing seasons. CVs for soybean yields ranged from low to moderate (9% to 16%). Therefore, temporal CV of soybean yield was 78% during these two growing seasons. Thus, crop yields varied spatially and temporally, as reported by previous studies (Jaynes and Colvin 1997; Lamb et al. 1997). This spatial-temporal variability was due to variations in growing season precipitation (Table 4-1). This represented a 25% variation during these four growing seasons.

4.6.2. Spatial structures of soil agri-environmental index (P/Al)M3 and other soil chemical and auxiliary properties

In the 0-5 cm layer, soil properties, including P indices were fitted with spherical models (Table 4-3). Spatial dependence ratios of soil P indices and other properties ranged from strong to moderate (43% to 99%). They were classified as follows: strong for soil pHwater and CaM3; strong-moderate for FeM3; and moderate for TC, AlM3 and soil P indices.

Spatial ranges varied from 55 to 77 m (Table 4-3). These values were higher than the 35-m grid spacing, indicating that the sampling grid designed for characterizing spatial variability of soil properties was appropriate. R^2 cv values ranged from 0.07 to 0.30 for soil properties and from 0.06 to 0.08 for soil P indices. Ranges varied among soil properties. The property with the smallest range will strongly influence required grid size (Lawrence et al. 2020), impacting the collection of grid samples for multiple soil properties (Metwally et al. 2019). Thus, the range of soil P indices was the smallest among soil properties and therefore controls the optimal sampling grid for soil P and other chemical properties.

Spatial dependence of soil P indices was moderate (43% to 46%). These results were lower than previous field-scale studies arising from high pedodiversity (McCormick et al. 2009; Miao et al. 2006; Nolin et al. 1999). Thus, pedodiversity may affect the spatial structure of soil P, impacting the delineation of MZs to optimize field-scale P fertilizer recommendations. A stronger spatial structure has positive impacts on delineating MZs, while a weaker spatial structure makes it difficult to use MZs because of a lack of spatial structure. Thus, a moderate spatial structure of soil P combined with use of ECa measurements would impact the relevance of MZDS for sustainable management of soil P in NT fields.

Auxiliary properties were fitted with spherical and exponential models (Table 4-3). Spatial dependence ratios of auxiliary properties ranged from strong to moderate (58% to 99%). They were classified as follows: moderate for yield2013; and strong for elevation, TWI, ECa measurements, and yield (2012, 2015 and 2016). Spatial ranges varied from 9.0 to 242 m. The highest ranges were observed by elevation, TWI, ECa measurements, and yield2015, indicating that these properties were influenced by extrinsic and natural factors over larger distances (López-Granados et al. 2002). R^2 cv values ranged from 0.25 to 0.97 for auxiliary properties and from 0.77 to 0.84 for ECa measurements.

Spatial dependencies of ECa measurements were strong (78% to 99%) (Table 4-3). This was due to (1) a strong influence of soil intrinsic factors, such as clay content (Cambardella et al. 1994), and (2) our sampling strategy density (229 ECa measurements per ha⁻ 1), as observed in previous field-scale studies (Cambouris et al. 2006; Moral et al. 2010). Perron et al. (2018) reported strong dependencies (R > 85%) for ECa measurements, corresponding to high sampling densities (394-452 samples ha¹). Thus, spatial mapping of ECa measurements will be performed with high reliability, aiming to delineate MZs. This was confirmed by the highest R^2 cv values (R^2 cv > 0.77) for ECa measurements.

Spatial dependencies of yields ranged from moderate to strong (58% to 99%) (Table 4-3). Similar results were reported by Rodrigues and Corá (2015) during five growing seasons under a 12-year NT field. Spatial dependencies of yields resulted from interactions between intrinsic and extrinsic soil factors (i.e. seasonal climate variability) (Cambouris et al. 2006). Thus, a moderate-strong dependence of yields expresses that yield variation was more impacted by spatial variability of soil properties than that induced by climate variability. R^2 cv values ranged from 0.25 to 0.56, meaning that spatial yield mapping can be performed reliably.

Thus, ECameasurements and yield (2012, 2015 and 2016) have the strongest spatial structures (77% to 99%). This means that these auxiliary properties are strongly spatially structured and controlled mainly by soil intrinsic properties (Ameer et al. 2022; Cambardella and Karlen 1999). Thus, spatial mapping of these auxiliary properties would be performed for delineating MZs, including ECa-based homogeneous MZs.

4.6.3. Spatial mapping of soil agri-environmental index (P/Al)M3 and other soil chemical and auxiliary properties

In the 0-5 cm NT soil layer, visual similarities were observed between PM3 and (P/A1)M3 spatial maps (Figs. 4-3 *a*, 3 *b*). Spatial values of these soil P indices (PM3 and (P/A1)M3) ranged from 30 to 120 mg kg⁻¹ and from 4.5% to 12.5%, respectively. Similar results were reported by Nze Memiaghe et al. (2021) from P accumulation in the 0-5 cm layer. The highest spatial values of soil P indices were generally observed in northern, southern and southwest field sections. The lowest spatial values of soil P indices were observed in centre and eastern field sections (Figs. *4-3a, 3b*). Thus, spatial mapping of soil P revealed development of large field-scale heterogeneity in the distribution pattern of soil P, as seen in previous studies (Bogunovic et al. 2014; Nze Memiaghe et al. 2021). Consequently, MZDS using $(P/Al)_{M3}$ measurements would represent a useful combination for accurate field-scale P recommendations, owing to this P spatial heterogeneity arising from this soil P variability (32% to 36%).

Spatial values of $A \mid M3$, TC, $_{FeM3}$ and $_{CaM3}$ ranged from 375 to 1150 mg kg^{-1} , from 14 to 27 g kg^{-1} , from 70 to 470 and from 1600 to 4400 mg kg^{-1} , respectively (Figs. *4-3c, 3d, 3e, 3f*). These results were lower for $_{AlM3}$ and TC and higher for $_{FeM3}$ and $_{CaM3}$ from previous studies in Eastern Canada (Nze Memiaghe et al. 2021; Quenum et al. 2012). Visual associations were generally observed between soil P indices and $_{CaM3}$ and TC, indicating their influence on soil P, as confirmed by results from Figure 4-5 and Table 4-4.

Spatial ECa values were lower in the 0-30 cm soil layer than the 0-100 cm soil layer (Figs. 4-4 *a*, 4 *b*). Indeed, $_{ECa30}$ values ranged from 10 to 30 mS m$^{\wedge 1}$, against 20 to 44 mS ιττ^1 for $_{ECa100}$ values (Figs. *4-4a, 4b*). This was because ECa is a measure that integrates many soil properties, varying with the depths of conductive soil materials (Doolittle et al. 1994; Kitchen et al. 2005). These spatial ECa values were higher than Nolin et al. (2003). The highest spatial ECa values were observed in the northern, southern and southwest field sections, whereas the lowest spatial ECa values were observed in the centre and eastern field sections (Figs. *4-4a, 4b*).

Spatial mapping of ECa measurements has agronomical implications for the delineation of ECa-based homogeneous MZs because it has the strongest spatial structures (78% to 99%) and highest R^2 cv values (r > 0.77). Additionally, ECa maps are important tools for delineating MZs because of their strong correlation with many soil properties (Corwin and Lesch 2003; Kitchen et al. 2003). Consequently, ECa measurements may represent a useful MZDS for predicting accurate P fertilizer recommendations. This may be dependent on the ECa-soil $(P/Al)_{M3}$ relationship.

Spatial values of corn yields were lower in 2012 than 2016, ranging from 6 to 13.5 Mg ha^{-1} against 13.5 to 16 Mg ha^{-1} , respectively (Figs. *4-4c, 4f*). Spatial values of soybean yields were lower in 2013 than 2015, ranging from 2.7 to 4.5 Mg ha^{-1} against 2.1 to 5 Mg ha^{-1} , respectively (Figs. *4-4d, 4e*). These increases in yield values were due to variation in climate conditions, as observed in Table 4-1. Increases in precipitation ranged from -15% to 6% and from 3% to 4% for both respective crops, compared to the 30-year normal. Similarly, increases in temperatures ranged from 7% to 6% and from 3% to 7% for both respective crops, compared to the 30-year normal (Table 4-1). Thus, yield variability was impacted by seasonal weather variation, as seen in previous similar studies (Miao et al. 2006; Nyéki et al. 2022; Rodrigues and Corá 2015).

4.6.4. Relationships between soil $(P/Al)_{M3}$, ECa measurements and other soil chemical and auxiliary properties

The highest correlations (r) were observed between $(P/Al)_{M3}$ and $_{CaM3}$, $_{pHwater}$, TC and $_{FeM3}$ (Fig. 4-5). Similar $(P/Al)_{m3-CaM3}$ and $(P/Al)_{M3-pHwater}$ correlations and lower $(P/A1)_{M3-TC}$ correlations were reported by previous studies (Metwally et al. 2019; Nolin et al. 1999). Correlations were observed between $_{ECa30}$ and $_{CaM3}$, $_{pHwater}$, P indices and $_{FeM3}$ (Fig. 4-5). $_{ECa30-CaM3}$ and $_{ECa30-pHwater}$ correlations were lower than previous studies (Bhunia et al. 2018; Cambouris et al. 2006; Perron et al. 2018). Significant ECa30-soil P indices correlations were similar to results from previous studies (Omonode and Vyn 2006; Singh et al. 2016).

Correlations (r) were also observed between $_{ECa100}$ and $_{CaM3}$, TC, and $_{pHwater}$ (Fig. 4-5). However, no $_{ECa100-P}$ indices correlation was observed, similarly to Peralta and Costa (2013). This was for two reasons: (I) chemically, equivalent conductances of common inorganic P ions (e.g. H2PO4$^-$, $_{HPO4^2}$ -) were generally lower than other ionic species (e.g. Ca^{2+} and Mg^{+2}) in soils; and (2) impacts of fertilization forms (band application) and tillage systems (direct drilling, without soil removal) on this low ECa-soil P relationship (Jung et al. 2005; Motavalli et al. 2015). Thus, further studies need to be performed to evaluate long-term impacts of contrasted tillage practices on the ECa-soil P relationship in Eastern Canada.

Moreover, $_{ECa100-soil}$ property correlation values were generally lower than ECa30-soil property correlation values (Fig. 4-5). This was similar to findings of Perron et al. (2018) and opposite to findings from previous studies (Lajili et al. 2021; Omonode and Vyn 2006; Singh et al. 2016). Thus, ECa-soil property relationships varied among fields because of the performance of field-specific ECa sensors when the field is dominated by one or two main intrinsic factors, such as clay content or soil moisture (Brevik et al. 2006; Corwin and Lesch 2005). These two intrinsic soil factors make interpretation of soil ECa values highly specific to the field.

Varying relationships between $(P/Al)_{M3}$, ECa measurements and yields (2012, 2015, and 2016) were observed because of climate variability (Fig. 4-5). Soil P-yield relationships varied among years (Guedes Filho et al. 2010; Nyéki et al. 2022) because of both dominant factors (i.e. soil properties and climate conditions), impacting yield variability (Jiang and Thelen, 2004; Nyéki et al. 2022; Singh et al. 2016). Varying ECa-yield relationships were reported by previous studies (Ameer et al. 2022; Nyéki et al. 2022). Crop yields were induced by climate variability, significantly impacting ECa measurements.

Spatial regression equations were generated to investigate the impacts of soil chemical and auxiliary properties on $(P/Al)_{M3}$ variability (Table 4-4). The highest regression values (R^2) were observed between $(P/Al)_{M3}$ and Ca_{M3}, pH_{water} and TC, ranging from 0.13 to 0.35. This indicated that 13% to 35% of $(P/Al)_{M3}$ agri- environmental variations were significantly explained by these soil properties. Thus, these spatial regression equations confirmed the strength and stability of relationships between $(P/Al)_{M3}$ and Ca_{M3}, pH_{water} and TC.

The lowest regression values (R^2) were observed between $(P/Al)_{M3}$ and ECa measurements and yields, ranging from 0.003 to 0.047 (Table 4-4). This indicated a low impact (0.3% to 5%) of these auxiliary properties on $(P/Al)_{M3}$ variability. Similar results were reported between soil P and ECa measurements under a 50-year corn-soybean rotation (Omonode and Vyn 2006). However, the highest R^2 values between soil P and ECa measurements may be affected by soil moisture from higher precipitations (Omonode and Vyn 2006; Rhoades et al. 1989). Thus, future studies need to be performed to evaluate soil moisture impacts on the ECa-soil P relationship under contrasted tillage practices in Eastern Canada.

Spatial information for delineating MZs should be quantitative, densely measured, structured, temporally stable, and closely related to crop yields (Doerge 2011). The CVs of $(P/Al)_{M3}$ measurements (32%) were higher than the CVs of ECa measurements (9% to 16%). Thus, delineation of P-based homogeneous MZs will be based on $(P/A1)_{M3}$ measurements as well as ECa measurements. Additionally, the ECa_{30}-$(P/A1)_{M3}$ Pearson's correlations were higher than the ECa_{100}-$(P/A1)_{M3}$ correlations. Consequently, three MZDS will be performed for reducing spatial variability of $(P/A1)_{M3}$, aiming to predict accurate P recommendations under the NT field. These MZDS will be based on (I) $(P/A1)_{M3}$ measurements; (2) ECa_{30} measurements; and (3) a combination of $(P/Al)_{M3+ECa30}$ measurements.

4.6.5. Delineation of homogeneous MZs using three strategies

Two to five MZs were delimited using $(P/A1)_{M3}$, ECa_{30} and $(P/Al)_{M3+ECa30}$ delineation strategies (Figs. *4-6a-61*). These MZs ranged from very low to very high. The mean $(P/A1)_{M3}$, ECa_{30} and $(P/Al)_{M3+ECa30}$ classes were (1) very low, (2) low, (3) medium, (4) high and (5) very high.

Decreasing visual similarities were observed from two to five MZs when $(P/Al)_{M3}$ and ECa_{30} delineation strategies were compared visually (Figs. *4-6a-6e, 6b-6f, 6c-6g* and *6d-6h)*. For two MZs, delineation of MZs seemed generally similar using both strategies. However, as the number of MZs increased, the delineation of MZs from these two strategies became increasingly different. Conversely, strong visual similarities between the ECa_{30} and $(P/Al)_{M3+ECa30}$ strategies were observed from two to five MZs (Figs. 4-6 e-6 *i*, 6 *f-6j*, 6 *g*-6 *k* and 6 *h - 6l*). Indeed, from two to five MZs, the lowest values were observed in the centre and eastern field sections, whereas the highest values were observed in the northern, southern and extreme western field sections.

Two main aspects were revealed from these results. First, these three MZDS successfully delineated MZs (2 to 5 MZs) within-field. Second, implementing the $(P/Al)_{M3}$ and $(P/Al)_{M3+ECa30}$ strategies represented new original MZDS for controlling spatial variability of soil P in the Eastern Canadian region. Indeed, delineation of MZs using ECa measurements was previously and successfully applied for many soil properties, including P in Eastern Canada (Cambouris et al. 2019; Lajili et al. 2021). Consequently $(P/Al)_{M3}$, ECa_{30} and $(P/Al)_{M3+ECa30}$ spatial maps may have potential for determining the optimum number of MZs to predict specific field-scale P recommendations.

4.6.6. Evaluation of three MZDS and determination of optimum number of MZs

1)   Delineation strategy using (P/Al)M3 and ECa30 measurements

Subdividing the field from one to five MZs decreased total variances of (P/Al)M3 and ECa30 from 100% to 10%, reducing their total variances from 73% and 90% (Figs. *4-7a, 7e*). However, increasing from one to two MZs reduced total within-zone variance for (P/Al)M3 and ECa30 from 100% to 38% (Figs. *4-7a, 7e*). These values were the highest reductions, representing 45% and 62% decreases of total within-zone variances (Figs. 4-7a, 7e). The inflexion point was reached from one to two MZs. Thus, two MZs was the optimal number for (P/Al)M3 and ECa30.

Subdividing the field from one to five MZs decreased total variances of ECa30 and (P/Al)M3 from 100% to 88%, reducing their total variances from 12% and 9% (Figs. *4-7b, 7f*). However, increasing from one to two MZs reduced total within-zone variance for ECa30 and (P/Al)M3 from 100% to 92%, representing 5% and 8% decreases of total within-zone variances (Figs. *4-7b, 7f*). Increasing from three to four MZs, represented a 10% decrease of total variance (Fig. *4-7b*). Delineating more than three MZs does not increase availability of spatial information (Peralta and Costa 2013).

Subdividing the field from one to five MZs decreased total variances of soil PM3, FeM3, TC, pHwater and CaM3 from 100% to 50%, reducing their total variances from 9% to 50% (Figs. 4-7c, 7g). However, increasing from one to two MZs had a greater impact than any other increase in MZ. Indeed, total within-zone variances for these soil properties were reduced from 100% to 70%, representing a 30% maximum decrease of total within- zone variance for these soil properties (Figs. *4-7c, 7g*). Thus, two MZs was the optimal number for soil PM3, FeM3, TC, pHwater and CaM3.

Subdividing the field from one to five MZs decreased total variances of yields (2012 to 2016) from 100% to 89%, reducing their total variances from 0% to 11% (Figs. *4-7d, 7h*). However, for yields (2015, 2016), increasing from one to two MZs had a slighter impact than any increase in MZ. Total within-zone variances for these auxiliary properties were reduced from 100% to 89%, representing 5% and 11% maximum decreases of their total within-zone variances (Figs. *4-7d, 7h*). Thus, two MZs was the optimal number for yields (2015, 2016).

2)   Delineation strategy using (P/Al)M3+ECa30 measurements

Subdividing the field from one to five MZs decreased (P/Al)M3 and ECa30 variances from 100% to 16%, reducing their total variances from 39% to 84% (Fig. *4-8a*). However, increasing from one to two MZs reduced their total within-zone variances from 100% to 42%, representing decreases from 16% to 58% of total within- zone variance (Fig. *4-8a*). The inflexion point was reached from one to two MZs. Thus, two MZs was the optimal number for (P/Al)M3 and ECa30.

Subdividing the field from one to five MZs decreased total variances of TC, FeM3, pHwater, CaM3 and PM3 from 100% to 63%, reducing their total variances from 4% to 37% (Fig. *4-8b*). However, increasing from one to two MZs had a greater impact than any other increase in MZ. Indeed, total within-zone variances for these soil properties were reduced from 100% to 71%, representing a 29% maximum decrease of total within-zone variance (Fig. *4-8b*). Thus, two MZs was the optimal number for soil FeM3, PM3, pHwater and CaM3.

Subdividing the field from one to five MZs decreased total variances of yields (2012 to 2016) from 100% to 88%, reducing their total variances from 0% to 12% respectively (Fig. *4-8c*). However, for yields (2015, 2016), increasing from one to two MZs had a slight impact in comparison with any other increase in MZ. Indeed, total within-zone variances for these auxiliary properties were reduced from 100% to 91%, representing a 9% maximum decrease of total within-zone variance (Fig. *4-8c*). Thus, two MZs represented the optimal number for yields (2015, 2016).

4.6.7. Practical applications of MZs within the site

1)   Statistical validations

Delineating two MZs resulted in significant differences of soil P indices, FeM3, pHwater and CaM3 values using three MZDS (Table 4-5). Thus, it seems statistically relevant to manage (P/Al)M3 and other soil properties by using three MZDS to develop two MZs. Similar results were reported by Nolin et al (2003). Delineating three MZs using three MZDS did not significantly impact different values of soil properties, except for soil P indices and CaM3, delineated using the (P/Al)M3 and ECa30 strategies, respectively (Table 4-5). Thus, it seems statistically relevant to manage (P/Al)M3 by using (P/Al)M3 measurements to develop three MZs in this study. Delineating three MZs for soil P using ECa measurements had results similar to previous studies (Lajili et al. 2021; Peralta and Costa 2013). Three or more MZs were not generally considered significant at the chosen 5% level (Perron et al. 2018).

Delineating two or three MZs resulted in significant differences of ECa30 values using the ECa30 and (P/Al)M3+ECa30 strategies (Table 4-5). Thus, it seems statistically relevant to manage ECa_{30} using both MZDS to develop two or three MZs. Similar results were reported recently (Lajili et al. 2021; Yuan et al. 2022). However, performance of three MZDS for developing two or three MZs had varying results on yield values for different years, probably because of climate variability (Table 4-5). A lack of significant yield difference between MZs results from low within-field spatial variability (Perron et al. 2018).

2)   P fertilizer recommendations and agri-environmental implications

Mean $(P/AI)_{M3}$ value was 7.9% at the 0-5 cm NT soil layer. Following local recommendations (CRAAQ 2010), P fertilizer applications were 40 kg and 0 kg P2O5 ha^{-1} for corn and soybean, respectively. Based on these values, local uniform P recommendations in the entire field were 380 kg and 0 kg P2O5 for corn and soybean, respectively (Table 4-6). This uniform agronomy application represents conventional agriculture (1MZ). Nevertheless, local P fertilizer recommendations were developed in Eastern Canada for uniform P agronomic applications regardless of field size, soil P variability or their crop productivity potential.

Variability of soil P was moderate (32% to 36%), indicating that existing uniform P fertilizer recommendations for corn and soybean cannot be suited to this large NT field (10 ha). By ignoring spatial variations in soil P fertility, P fertilizer applications could result in P over-fertilization and under-fertilization in large fields (Schumann et al. 2003; Yuan et al. 2022). Consequently, accurate P fertilizer recommendations were developed for both crops, taking into account (1) spatial values for soil P from soil P variability and (2) corresponding Quebec agronomic P recommendations for these crops following local recommendations (CRAAQ 2010). Thus, delineating two and three MZs using these three MZDS generated six accurate P fertilizer recommendations for corn and soybean. These P fertilizer values were compared to P fertilizer recommendations from uniform application (i.e. conventional method) (Table 4-6).

Thus, delineating two and three MZs using strategy 1 generated a total of 306 and 341 kg P2O5, respectively, for corn and 0 kg P2O5 for soybean (Table 4-6). Uniform P recommendations (1MZ) were 380 kg and none (0 kg P2O5) for corn and soybean, respectively (Table 4-6). This represented 74 kg and 40 kg P2O5 reductions for two and three MZs, respectively, for corn and none (0 kg P2O5) for soybean. Delineating two and three MZs using other strategies (2 and 3) generated 380 kg P2O5 and 0 kg P2O5 for corn and soybean, respectively (Table 4-6). These P fertilizer recommendations were similar to uniform P recommendations (380 kg and 0 kg P2O5) for both crops (Table 4-6). Therefore, there were no P2O5 fertilizer reductions for two and three MZs using both strategies (2 and 3), respectively, for both crops compared to uniform P recommendations.

Thus, our results suggested that delineating two or three MZs using the $(P/AI)_{M3}$ measurement represented the best agronomic strategy for this site-specific P management. This was due to (1) the largest reduction (45%) of spatial variability of $(P/AI)_{M3}$ occurred from one to two MZs, as confirmed by previous statistical validations (2-3 MZs), and (2) largest P2O5 reductions from two or three MZs (74 kg P2O5 or 8 kg ha^{-1} and 40 kg P2O5 or 4 kg ha^{-1}, respectively) compared with the other MZDS, including conventional agriculture. Many studies successfully delimited two optimal MZs (Nolin et al. 2003; Peralta and Costa 2013), while others found three optimal MZs (Rodrigues and Corá 2015; Yuan et al. 2022) associated with spatial distribution of soil P. Previous authors saved approximately 15-20 kg P ha^{-1} under large fields by delineating MZs compared to conventional agriculture (Moharana et al. 2020; Yuan et al. 2022).

Thus, there is a potential for reducing spatial variability of soil P by delineating two or three MZs to reduce P losses while maintaining yields in Eastern Canada. The agri-environmental $(P/AI)_{M3}$ index could be managed differently in each MZ. Moreover, development of MZs must be simple, functional and economically feasible for farmers to adopt site-specific management (Li et al. 2007; Nolin et al. 2003). The choice of number of MZs will depend on the degree of precision a farmer wants to reach and on the amount of money the farmer is ready to invest in variable-rate technologies.

4.7. Conclusions

To our knowledge, this study represents the first on delineating MZs to control the spatial variability of soil P under a NT field in Eastern Canada. The main goal was to delineate MZs using the ECa-P relationship to reduce soil sampling and spatial variability of $(P/AI)_{M3}$ in order to develop accurate P recommendations under NT fields in Eastern Canada. Results revealed that the mean $(P/AI)_{M3}$ value was 7.9% at the 0-5 cm NT soil layer. The mean values of ECa_{30} and ECa_{100} were 15.8 and 32.6 mS m^{-1}, respectively. The CV of soil P indices remained high (32%

to 36%) despite a low pedodiversity within this field. Spatial dependence of soil P and auxiliary properties ranged from moderate to strong (43% to 99%). A significant relative ECa$_{30}$ -soil P correlation was observed, generating three MZDS.

Variability of soil P was moderate, indicating that existing uniform P fertilizer recommendation is not suited to this large NT field (10 ha). Thus, there is the potential to reduce spatial variability of soil P using MZDS in order to reduce P losses in Eastern Canada. In keeping with results of this study, delineating two or three MZs using (P/Al)$_{M3}$ measurements represented the best agronomic strategy under this NT field. This was due to largest reduction (45%) of spatial variability of (P/A1)$_{M3}$, as confirmed by statistical validations (2-3 MZs), and the largest P2O5 reductions from two or three MZs (74 kg P2O5 or 40 kg P2O5, respectively).

This study was performed under one selected NT field. Further studies in other fields are required to evaluate the long-term impacts of contrasted tillage practices and soil moisture content on the ECa-soil P relationship in Eastern Canadian soils. This will aim to predict spatial distribution of soil P agri-environmental index (P/Al)$_{M3}$ for improving P fertilization recommendations. Likewise, future studies are required to take into account temporal variability of soil P in relationship with temporal variability of crop yields to delineate temporal MZs accurately. Soil properties and climate conditions may vary each year in the context of climate variation. The optimum number of MZs can change over time and is a function of the weather and the crop planted. Thus, future choices of MZ numbers should be based in the temporal data set.

4.8. Acknowledgements

This research project was funded by Agriculture and Agri-food Canada (AAFC), grant number J- 002238. We would like to thank Claude Lévesque and Sylvie Michaud for their field assistance and technical support.

4.9. References

Abdi, D., Cade-Menun, B.J., Ziadi, N., and Parent, L.E. 2014. Long-term impact of tillage practices and phosphorus fertilization on soil phosphorus forms as determined by 31 P nuclear magnetic resonance spectroscopy. J. Environ. Qual. **43**(4): 1431-1441. doi :http://dx.doi.org/10.2134/jeq2013.10.0424.

Adamchuk, V., Allred, B., Doolittle, J., Grote, K., and Rossel, R. 2015. Tools for proximal soil sensing. in Soil Survey Staff, C. Ditzler, and L. West, eds. Soil survey manual. United States Department of Agriculture Handbook No. 18, Natural Resources Conservation Service. [Online]. Available from https://www.nrcs.usda.gov/wps/portal/nrcs/detail/soils/ref/? cid=nrcs142p2 054256.

Ameer, S., Cheema, M.J.M., Khan, M.A., Amjad, M., Noor, M., and Wei, L. 2022. Delineation of nutrient management zones for precise fertilizer management in wheat crop using geo-statistical techniques. Soil Use Manag. 38 (3): 1430-1445. doi :https://doi.org/10.1111/sum.12813.

Argento, F., Anken, T., Abt, F., Vogelsanger, E., Walter, A., and Liebisch, F. 2021. Site-specific nitrogen management in winter wheat supported by low-altitude remote sensing and soil data. Precis. Agric.**22**: 364-386. https://doi.org/10.1007/s11119-020-09733-3.

Armstrong, R., Fitzpatrick, J., Rab, M., Abuzar, M., Fisher, P., and O'leary, G.J. 2009. Advances in precision agriculture in south-eastern Australia. III. Interactions between soil properties and water use help explain spatial variability of crop production in the Victorian Mallee. Crop Pasture Sci. **60**(9): 870-884. doi: https://doi.org/10.1071/CP08349.

Becker, S.M., Franz, T.E., Abimbola, O., Steele, D.D., Flores, J.P., Jia, X., Scherer, T.F., Rudnick, D.R., and Neale, C. M.2022. Feasibility assessment on use of proximal geophysical sensors to support precision management. Vadose Zone J. **21**(6). doi: https://doi.org/10.1002/vzj2.20228.

Beven, K.J., and Kirkby, M. 1979. A physically based, variable contributing area model of basin hydrology. Hydrol. Sci. J. **24** (1): 43-69. doi: https://doi.org/10.1080/02626667909491834.

Bhunia, G.S., Shit, P.K., and Chattopadhyay, R. J. 2018. Assessment of spatial variability of soil properties using geostatistical approach of lateritic soil (West Bengal, India). Ann. Agrar. Sci. **16** (4): 436-443. doi : https://doi.org/10.1016/j.aasci.2018.06.003.

Bogunovic, I., Mesic, M., Zgorelec, Z., Jurisic, A., and Bilandzija, D. 2014. Spatial variation of soil nutrients on sandy-loam soil. Soil Till. Res. **144**: 174-183. doi :10.1016/j.still.2014.07.020.

Bolinder, M., Simard, R., Beauchemin, S., and Macdonald, K.B. 2000. Indicator of risk of water contamination by

P for soil landscape of Canada polygons. Can. J. Soil Sci. **80**(1): 153-163. doi: https://cdnsciencepub.com/doi/pdf/10.4141/S99-040.

Bongiovanni, R., and Lowenberg-Deboer, J. 1999. Economics of variable rate lime in Indiana. Proceedings of the fourth international conference on precision agriculture. Wiley Online Library.pp. 1653-1665. doi: 10.2134/1999.precisionagproc4.c70b.

Borges, R., and Mallarino, A. (1998). Significance of spatially variable soil phosphorus and potassium for early growth and nutrient content of no-till corn and soybean. Commun. Soil Sci. Plant Anal. **29** (17-18): 2589-2605.

Brevik, E.C., Fenton, T.E., and Lazari, A. 2006. Soil electrical conductivity as a function of soil water content and implications for soil mapping. Precis.   Agric.7:393-404. doi: https://link.springer.com/article/10.1007/s11119-006-9021-x.

Brock, A., Brouder, S., Blumhoff, G., and Hofmann, B.S. 2005. Defining yield-based management zones for corn-soybean rotations. Agron. J. : **97**(4): 1115-1128. doi: https://doi.org/10.2134/agronj2004.0220.

Brune, D., and Doolittle, J. 1990. Locating lagoon seepage with radar and electromagnetic survey. Environ. Geol. **16**: 195-207. doi: https://link.springer.com/article/10.1007/BF01706044.

Bullock, D.S., and Bullock, D.G. 2000. From agronomic research to farm management guidelines: A primer on the economics of information and precision technology. Precis. Agric. **2**: 71-101. doi: https://link.springer.com/article/10.1023/A: 1009988617622.

Cade-Menun, B.J., Carter, M.R., James, D.C., and Liu, C.W. 2010. Phosphorus forms and chemistry in the soil profile under long-term conservation tillage: A phosphorus-31 nuclear magnetic resonance study. J. Environ. Qual. **39**(5):1647-1656. doi: https://doi.org/10.2134/jeq2009.0491.

Cambardella, C., and Karlen, D.L. 1999. Spatial analysis of soil fertility parameters. Precis. Agric.**1** (1): 5-14. doi: https://link.springer.com/article/10.1023/A:1009925919134.

Cambardella, C., Moorman, T., Parkin, T., Karlen, D., Novak, J., Turco, R., and Konopka, A. 1994. Field-scale variability of soil properties in central Iowa soils. Soil Sci. Soc. Am. J. **58**(5): 1501-1511. doi: https://doi.org/10.2136/sssaj1994.03615995005800050033x.

Cambouris, A., Messiga, A., Ziadi, N., Perron, I., and Morel, C. 2017. Decimetric-Scale Two-Dimensional Distribution of Soil Phosphorus after 20 Years of Tillage Management and Maintenance Phosphorus Fertilization. Soil Sci. Soc. Am. J. **81**(6): 1606-1614. doi: https://doi.org/10.2136/sssaj2017.03.0101.

Cambouris, A.N., Nolin, M.C., Zebarth, B.J., and Laverdière, M.R. 2006. Soil management zones delineated by electrical conductivity to characterize spatial and temporal variations in potato yield and in soil properties. Am. J. Potato Res.**83**(5):381-395.doi: https://link.springer.com/article/10.1007/BF02872015.

Cambouris, A.N., Nolin, M.C., and Simard, R.R. 1999. Precision management of fertilizer phosphorus and potassium for potato in Quebec, Canada. Proc. 4th International Conference on Precision Agriculture. ASA-CSA-SSSA, Madison, WI, USA. pp. 847-857.

Cambouris, A.N., Zebarth, B.J., and Perron, I. 2019. Delineated Soil Management Zones with Proximal Soil Sensors to Effectively Characterize the Soil Variability under Potato Production in Eastern Canada. 5th Global Workshop on Proximal Soil Sensing. Columbia, Missouri, USA. pp. 9-14.

Cambouris, A.N., Zebarth, B.J., Ziadi, N., and Perron, I. 2014. Precision agriculture in potato production. Potato Res. **57**: 249-262. doi: https://link.springer.com/article/10.1007/s11540-014-9266-0.

Cassidy, R., Thomas, I.A., Higgins, A., Bailey, J.S., and Jordan, P. 2019. A carrying capacity framework for soil phosphorus and hydrological sensitivity from farm to catchment scales. Sci. Total. Environ. **687**: 277- 286. doi: https://doi.org/10.1016/j.scitotenv.2019.05.453.

Corwin, D., and Lesch, S.M. 2003. Application of soil electrical conductivity to precision agriculture: theory, principles, and guidelines. Agron. J. : **95**(3): 455-471. doi: https://doi.org/10.2134/agronj2003.4550.

Corwin, D.L., and Lesch, S.M. 2005. Apparent soil electrical conductivity measurements in agriculture. Comput. Electron. Agric. **46** (1-3): 11-43. doi: https://doi.org/10.1016/j.compag.2004.10.005.

Cox, M., Gerard, P.D., Wardlaw, M., and Abshire, M.J. 2003. Variability of selected soil properties and their relationships with soybean yield. Soil Sci. Soc. Am. J. **67**(4)**:** 1296-1302. doi : https://doi.org/10.2136/sssaj2003.1296.

CRAAQ. 2010. Guide de référence en fertilisation. 2nd edition. Quebec: Centre de Référence en Agriculture et Agroalimentaire du Québec. QC, Canada. p. 473.

Doerge, T.A. 2011. Management zone concepts. Site-specific management guidelines. SSMG-2. International Plant Nutrition Institute, Norcross, GA. 4 p.

Doerge, T.A. 1999. Management zone concepts. SSMG-2. Potash & Phosphate Institute. International Plant NutritionInstitute , Norcross, GA. doi: http://www.ipni.net/publication/ssmg.nsf/0/C0D052F04A53E0BF852579E500761AE3/$FILE/SSMG- 02.pdf.

Doolittle, J., Sudduth, K., Kitchen, N., Indorante. 1994. Estimating depths to claypans using electromagnetic induction methods. J Soil Water Conserv. 49(6):572-575. doi: https://www.ars.usda.gov/ARSUserFiles/50701000/cswq-0270-doolittle.pdf

Environment Canada. 2013. Canadian Climate Normals or Averages 1971-2000 [Online]. Available from: http://www.climat.meteo.ec.gc.ca/climate normals/index e.html [Accessed 27 November 2022].

Ferguson, R., Lark, R., and Slater, G. P. 2003. Approaches to management zone definition for use of nitrification inhibitors. Soil Sci. Soc. Am. J. **67**(3) 937-947. doi: https://doi.org/10.2136/sssaj2003.9370.

Fraisse, C., Sudduth, K., and Kitchen, N.R. 2001. Delineation of site-specific management zones by unsupervised classification of topographic attributes and soil electrical conductivity. Trans. ASABE. **44**(1)155-166. doi: 10.13031/2013.2296.

Goovaerts, P. 1998. Geostatistical tools for characterizing the spatial variability of microbiological and physicochemical soil properties. Biol. Fertil. Soils **27:** 315-334. doi: https://doi.org/10.1007/s003740050439.

Guastaferro, F., Castrignanò, A., De Benedetto, D., Sollitto, D., Troccoli, A., and Cafarelli, B. 2010. A comparison of different algorithms for the delineation of management zones. Precis. Agric.**11:** 600-620. doi: https://link.springer.com/article/10.1007/s11119-010-9183-4.

Guedes Filho, O., Vieira, S.R., Chiba, M.K., Nagumo, C.H., and Dechen, S.C.F. 2010. Spatial and temporal variability of crop yield and some Rhodic Hapludox properties under no-tillage. Rev. Bras. Cienc. Solo **34:** 1-14. doi: https://link.springer.com/article/10.1007/s11119-010-9183-4.

Guo-Shun, L., Xin-Zhong, W., Zheng-Yang, Z., and Chun-Hua, Z. 2008. Spatial variability of soil properties in a tobacco field of central China. Soil Science **173**(9):659-667.doi: https://journals.lww.com/soilsci/Fulltext/2008/09000/SPATIAL VARIABILITY OF SOIL PROPERTIE S IN A.7.aspx.

Haghverdi, A., Leib, B.G., Washington-Allen, R.A., Ayers, P.D., and Buschermohle, M.J. 2015. Perspectives on delineating management zones for variable rate irrigation. Comput. Electron. Agric. 117:154-167. doi: https://www.sciencedirect.com/science/article/pii/S0168169915002264?via%3Dihub.

Han, S., Evans, R.G., Schneider, S.M., and Rawlins, S.L. 1996. Spatial variability of soil properties on two center pivot irrigated fields. Proceedings of the Third International Conference on Precision Agriculture. American Society of Agronomy, Crop Science Society of America, Soil Science Society of America. Madison, WI, USA. pp. 95-106.

Heilig, J., Kempenich, J., Doolittle, J., Brevik, E.C., and Ulmer, M. 2011. Evaluation of electromagnetic induction to characterize and map sodium-affected soils in the Northern Great Plains. Soil Survey Horizons **52**(3): 77-88. doi: https://acsess.onlinelibrary.wiley.com/doi/10.2136/sh2011.3.0077.

Heiniger, R.W., Mcbride, R.G., and Clay, D.E. 2003. Using soil electrical conductivity to improve nutrient management. Agron. J. **95**(3) 508-519. doi: https://acsess.onlinelibrary.wiley.com/doi/10.2134/agronj2003.5080.

Hendershot, W.H., Lalande, H., and Duquette, M. 2008. Soil reaction and exchangeable acidity. Pages 173-178 in R. Carter and E.G. Gregorich, eds. Soil sampling and methods of analysis. Taylor & Francis, Boca Raton, FL, USA.

Higgins, S., Schellberg, J., and Bailey, J. 2019. Improving productivity and increasing the efficiency of soil nutrient management on grassland farms in the UK and Ireland using precision agriculture technology. Eur. J. Agron. **106:** 67-74. doi: https://doi.org/10.1016/j.eja.2019.04.001.

Holm, F., Zentner, R., Thomas, A., Sapsford, K., Légère, A., Gossen, B., Olfert, O., and Leeson, J. 2006. Agronomic and economic responses to integrated weed management systems and fungicide in a wheat-canola-barley-pea rotation. Can. J. Plant Sci. **86**(4): 1281-1295.   doi : https://doi.org/10.4141/P05-165.

Jaynes, D.B., and Colvin, T.S. (1997). Spatiotemporal variability of corn and soybean yield. Agron. J. **89** (1): 3037. doi: https://doi.org/10.2134/agronj1997.00021962008900010005x.

Jiang, H.-L., Liu, G.-S., Liu, S.-D., Li, E.-H., Wang, R., Yang, Y.-F., and Hu, H.-C. 2012. Delineation of sitespecific management zones based on soil properties for a hillside field in central China. Arch. Agron. Soil Sci. **58** (10):1075-1090. doi: https://doi.org/10.1080/03650340.2011.570337.

Jiang, P., and Thelen, K.D. 2004. Effect of soil and topographic properties on crop yield in a North-Central corn soybean cropping system. Agron. J. **96** (1): 252-258. doi: https://doi.org/10.2134/agronj2004.0252.

Jung, W., Kitchen, N., Sudduth, K.A., Kremer, R., and Motavalli, P. 2005. Relationship of apparent soil electrical conductivity to claypan soil properties. Soil Sci. Soc. Am. J. **69**(3): 883-892. doi: https://doi.org/10.2136/sssaj2004.0202.

Kassam, A., Friedrich, T., Derpsch, R., Lahmar, R., Mrabet, R., Basch, G., González-Sánchez, E.J., and Serraj, R. 2012. Conservation agriculture in the dry Mediterranean climate. Field Crops Res. **132**: 7-17. doi: https://doi.org/10.1016/j.fcr.2012.02.023.

Khiari, L., Parent, L., Pellerin, A., Alimi, A., Tremblay, C., Simard, R., and Fortin, J. 2000. An agri-environmental phosphorus saturation index for acid coarse-textured soils. J. Environ. Qual.**29**(5):1561-1567. American Society of Agronomy, Crop Science Society of America, Soil Science Society of America. Madison, WI, USA. pp. 1561-1567. doi: https://doi.org/10.2134/jeq2000.00472425002900050024x.

Khosla, R., Fleming, K., Delgado, J., Shaver, T., and Westfall, D.G. 2002. Use of site-specific management zones to improve nitrogen management for precision agriculture. J Soil Water Conserv. **57**(6): 513518. doi: https://www.jswconline.org/content/57/6/513.short.

Khosla, R., and Shaver, T. 2001. Zoning in on nitrogen needs. Colorado State University Agronomy Newsletter **21**(1): 24-26.

Kitchen, N., Drummond, S., Lund, E., Sudduth, K., and Buchleiter, G. 2003. Soil electrical conductivity and topography related to yield for three contrasting soil-crop systems. Agron. J. **95**(3): 483-495. doi: https://doi.org/10.2134/agronj2003.4830.

Kitchen, N., Sudduth, K., Myers, D., Drummond, S., and Hong, S.Y. 2005. Delineating productivity zones on claypan soil fields using apparent soil electrical conductivity. Comput. Electron. Agric. 46(1-3): 285308. doi: https://10.1016/j.compag.2004.11.012.

Kravchenko, A., Bollero, G.A., Omonode, R., and Bullock, D. 2002. Quantitative mapping of soil drainage classes using topographical data and soil electrical conductivity. Soil Sci. Soc. Am. J. 66 **(1):** 235-243. doi: https://doi.org/10.2136/sssaj2002.2350.

Kweon, G., Lund, E., and Maxton, C. 2012. The ultimate soil survey in one pass: soil texture, organic matter, pH, elevation, slope, and curvature. Proceedings of the 11th International Conference on Precision Agriculture. ASA-CSA-SSSA, Madison, WI, USA. pp1-13.

Lafond, G.P., Walley, F., May, W., and Holzapfel, C. 2011. Long term impact of no-till on soil properties and crop productivity on the Canadian prairies. Soil Till. Res. **117**:   110-123.   doi: https://doi.org/10.1016/j.still.2011.09.006.

Lajili, A., Cambouris, A.N., Chokmani, K., Duchemin, M., Perron, I., Zebarth, B.J., Biswas, A., and Adamchuk, V. 2021. Analysis of four delineation methods to identify potential management zones in a commercial potato field in eastern Canada. Agronomy **11**(3) 432. doi: https://doi.org/10.3390/agronomy11030432.

Lamb, J., Dowdy, R., Anderson, J., and Rehm, G.W.1997. Spatial and temporal stability of corn grain yields. J. Prod. Agric. **10**(3): 410-414.

Lawrence, P.G., Roper, W., Morris, T.F., and Guillard, K. 2020. Guiding soil sampling strategies using classical and spatial statistics: A review. Agron. J. **112**(1): 493-510. doi: https://doi.org/10.1002/agj2.20048.

Li, Y., Shi, Z., Li, F., Li, H.-Y. 2007. Delineation of site-specific management zones using fuzzy clustering analysis in a coastal saline land. Comput. Electron. Agric. **56**(2):   174-186. doi: https://doi.org/10.1016/j.compag.2007.01.013.

Li, Y., Shi, Z., Wu, C.-F., Li, H.-Y., and Li, F. 2008. Determination of potential management zones from soil electrical conductivity, yield and crop data. J. Zhejiang Univ. Sci. **9**: 68-76. doi: https://link.springer.com/article/10.1631/jzus.B071379.

López-Granados, F., Jurado-Expósito, M., Atenciano, S., García-Ferrer, A., Sánchez De La Orden, M., and García-Torres, L. 2002. Spatial variability of agricultural soil parameters in southern Spain. Plant Soil **246**: 97-105. doi: https://l ink.springer.com/article/10.1023/A: 1021568415380.

Lund, E., Christy, C., and Drummond, P.E. 1999. Practical applications of soil electrical conductivity mapping. Precis. Agric. **99**: 771-779.

Machado, S., Bynum, E., Archer, T., Lascano, R., Wilson, L., Bordovsky, J., Segarra, E., Bronson, K., Nesmith, D., and Xu, W. 2002. Spatial and temporal variability of corn growth and grain yield: Implications for site-specific farming. Crop Sci. **42**(5):1564-1576. doi: https://d0i.0rg/l 0.2135/cropsci2002.1564.

Mallarino, A.P. 1996. Spatial variability patterns of phosphorus and potassium in no-tilled soils for two sampling scales. Soil Sci. Soc. Am. J.   **60**(5):1473-1481. doi: https://doi.org/10.2136/sssaj1996.03615995006000050027x.

Malvezi, K.E.D., Júnior, L.a.Z., Guimarães, E.C., Vieira, S.R., and Pereira, N. 2019. Soil chemical attributes variability under tillage and no-tillage in a long-term experiment in southern Brazil. Bioscience J. 35(2): 467-476. doi : 10.14393/BJ-v35n2a2019-41793.

MAPAQ. 2020. Portrait-Diagnostic sectoriel de l'industrie des grains au Québec. Ministère de l'Agriculture de l'Alimentation et des Pêcheries du Québec. p. 27.

Mccormick, S., Jordan, C., and Bailey, J. 2009. Within and between-field spatial variation in soil phosphorus in permanent grassland. Precis. Agric. **10**: 262-276. doi: https://link.springer.com/article/10.1007/s11119- 008-9099-4.

Messiga, A.J., Ziadi, N., Morel, C., and Parent, L.-E. 2010. Soil phosphorus availability in no-till versus conventional tillage following freezing and thawing cycles. Can. J. Soil Sci. **90**(3): 419-428. doi: https://doi.org/10.4141/CJSS09029.

Metwally, M.S., Shaddad, S.M., Liu, M., Yao, R.-J., Abdo, A.I., Li, P., Jiao, J., and Chen, X. 2019. Soil properties spatial variability and delineation of site-specific management zones based on soil fertility using fuzzy clustering in a hilly field in Jianyang, Sichuan, China. Sustainability 11(24): 7084. doi: https://doi.org/10.3390/su11247084.

Miao, Y., Mulla, D., and Robert, P. 2006. Spatial variability of soil properties, corn quality and yield in two Illinois, USA fields: implications for precision corn management. Precis. Agric. **7**(1): 5-20. doi: https://link.springer.com/article/10.1007/s11119-005-6786-2.

Moharana, P., Jena, R., Pradhan, U., Nogiya, M., Tailor, B., Singh, R., and Singh, S. 2020. Geostatistical and fuzzy clustering approach for delineation of site-specific management zones and yield-limiting factors in irrigated hot arid environment of India. Precis. Agric. **21**: 426-448. doi: https://link.springer.com/article/10.1007/s11119-019-09671-9.

Moral, F., Terrón, J., and Da Silva, J.M. 2010. Delineation of management zones using mobile measurements of soil apparent electrical conductivity and multivariate geostatistical techniques. Soil Till. Res. **106**(2): 335-343. doi: https://doi.orq/10.1016/j.still.2009.12.002-

Motavalli, P., Hammer, R., and Bardhan, S. 2015. Apparent soil electrical conductivity used to determine soil phosphorus variability in poultry litter-amended pastures. Yale Rev. Educ. Sci. 287-309.

Nolin, M., and Caillier, M. 1992. La variabilité des sols. II-Quantification et amplitude. Agrosol **5**: 21-32.

Nolin, M., Gagnon, B., Leclerc, M., Cambouris, A., Bélanger, G., and Simard, R. 2002. Influence of pedodiversity and past land uses on the within-field spatial variability of selected soil and forage quality indicators. Proceedings of the 6th International Conference on Precision Agriculture and Other Precision Resources Management. American Society of Agronomy, Minneapolis, MN, USA. pp. 261-277.

Nolin, M., Guertin, S., and Wang, C. 1996. Within-Field Spatial Variability of Soil Nutrients and Corn Yield in a Montreal Lowlands Clay Soil. Proceedings of the Third International Conference on Precision Agriculture. ASA-CSA-SSSA, Madison, WI, USA. pp. 257-270.

Nolin, M.C., Simard, R., Cambouris, A., and Beauchemin, S. 1999. Specific Variability of Phosphorus Status and Sorption Characteristics in Clay Soils of the St-Lawrence Lowlands (Quebec). Proceedings of the Fourth International Conference on Precision Agriculture. ASA-CSA-SSSA, Madison, WI, USA. pp. 395-406.

Nyéki, A., Daróczy, B., Kerepesi, C., Neményi, M., and Kovács, A.J. 2022. Spatial Variability of Soil Properties and Its Effect on Maize Yields within Field-A Case Study in Hungary. Agronomy **12**(2): 395. doi: https://doi.org/10.3390/agronomy12020395.

Nze Memiaghe, J.D., Cambouris, A.N., Ziadi, N., Karam, A., and Perron, I. 2021. Spatial variability of soil phosphorus indices under two contrasting grassland fields in Eastern Canada. Agronomy **11**(1): 24. doi: https://doi.org/10.3390/agronomy11010024.

Odeh, I.O., Mcbratney, A., and Chittleborough, D. 1995. Further results on prediction of soil properties from terrain attributes: heterotopic cokriging and regression-kriging. Geoderma **67**(3-4): 215-226. doi: 10.1016/0016-7061(95)00007-B.

Omonode, R.A., and Vyn, T.J. 2006. Spatial dependence and relationships of electrical conductivity to soil organic matter, phosphorus, and potassium. Soil Science **171**(3)   223-238.   doi: 10.1097/01.ss.0000199698.94203.a4.

Patoine, M., Hébert, S., Simoneau, M., and D'auteuil-Potvin, F. 2017. Loads of phosphorus, nitrogen and suspended solids at the mouths of Quebec rivers, 2009 to 2012. Ministère du Développement durable, de l[1] Environnement et de la Lutte contre les changements climatiques, Direction générale du suivi de l'état de l'environnement. p.25.

Pellerin, A., Parent, L.-É., Fortin, J., Tremblay, C., Khiari, L., and Giroux, M. 2006. Environmental Mehlich-III soil phosphorus saturation indices for Quebec acid to near neutral mineral soils varying in texture and genesis. Can. J. Soil Sci. **86**(4):711-723. doi: https://doi.org/10.4141/S05-070.

Peralta, N.R., and Costa, J.L. 2013. Delineation of management zones with soil apparent electrical conductivity to improve nutrient management. Comput. Electron. Agric. **99**: 218-226. doi: 10.1016/j.compag.2013.09.014.

Perron, I., Cambouris, A.N., Chokmani, K., Vargas Gutierrez, M.F., Zebarth, B.J., Moreau, G., Biswas, A., and Adamchuk, V. 2018. Delineating soil management zones using a proximal soil sensing system in two commercial potato fields in New Brunswick, Canada. Can. J. Soil Sci. 98**(4)** 724-737. doi: https://doi.org/10.1139/cjss-2018-0063.

Pierce, F.J., and Nowak, P. 1999. Aspects of precision agriculture. Adv. Agron. **67**: 1-85. doi: 10.1016/S0065-2113(08)60513-1.

Puustinen, M., Koskiaho, J., and Peltonen, K. 2005. Influence of cultivation methods on suspended solids and phosphorus concentrations in surface runoff on clayey sloped fields in boreal climate. Agric Ecosyst Environ. **105**(4):565-579. doi: 10.1016/j.agee.2004.08.005.

Quenum, M., Nolin, M., and Bernier, M. (2012). Digital mapping of the maximum phosphorus sorption capacity of soils at the agricultural plot scale using auxiliary variables. Can. J. Soil Sci. **92**(5): 733-750. doi: https://doi.org/10.4141/cjss2011-087.

Rab, M., Fisher, P., Armstrong, R., Abuzar, M., Robinson, N., Chandra, S.J.C., and Science, P. (2009). Advances in precision agriculture in south-eastern Australia. IV. Spatial variability in plant-available water capacity of soil and its relationship with yield in site-specific management zones. Crop Pasture Sci. **60**(9): 885-900. doi: https://doi.org/10.1071/CP08350.

Rhoades, J., Manteghi, N., Shouse, P., and Alves, W.J. 1989. Soil electrical conductivity and soil salinity: New formulations and calibrations. Soil Sci. Soc. Am. J. **53**(2):   433-439.   doi: https://doi.org/10.2136/sssaj1989.03615995005300020020x.

Robertson, G. 2008. GS+: Geostatistics for the Environmental Sciences. Gamma Design Software, Plainwell, MI.

Rodrigues, M.S., and Corá, J.E. 2015. Management zones using fuzzy clustering based on spatial-temporal variability of soil and corn yield. Eng. Agricola **35**:470-483. doi: http://dx.doi.org/10.1590Z1809-4430-Eng.Agric.v35n3p470-483/2015.

Saifuzzaman, M., Adamchuk, V., Biswas, A., and Rabe, N. 2021. High-density proximal soil sensing data and topographic derivatives to characterise field variability. Biosyst. Eng. **211**:   19-34. doi: https://doi.orq/10.1016/j.biosystemsenq.2021.08.018.

Sanches, G.M., Magalhães, P.S., Remacre, A.Z., and Franco, H.C. 2018. Potential of apparent soil electrical conductivity to describe the soil pH and improve lime application in a clayey soil. Soil Till. Res. **175**: 217-225. doi: https://doi.org/10.1016/j.still.2017.09.010.

Sas Institute. 2010. SAS user's guide. Statistics. Version 9.3. SAS Inst., Cary, NC. USA. SAS Institute.

Schepers, A.R., Shanahan, J.F., Liebig, M.A., Schepers, J.S., Johnson, S.H., and Luchiari Jr, A. 2004. Appropriateness of management zones for characterizing spatial variability of soil properties and irrigated corn yields across years. Agron J. **96**(1): 195-203. doi: 10.2134/agronj2004.0195.

Schumann, A.W., Fares, A., Alva, A.K., and Paramasivam, S. 2003. Response of Hamlin'orange to fertilizer source, annual rate and irrigated area. Proceedings of the Florida State Horticultural Society. pp. 256260.

Singh, G., Williard, K.W., and Schoonover, J.E. 2016. Spatial relationship of apparent soil electrical conductivity with crop yields and soil properties at different topographic positions in a small agricultural watershed. Agronomy 6(4): 57. doi: https://doi.org/10.3390/agronomy6040057.

SISCAN.1998. The Canadian system of soil classification. NRC Research Press.

Soane, B.D., Ball, B.C., Arvidsson, J., Basch, G., Moreno, F., and Roger-Estrade, J. 2012. No-till in northern, western and south-western Europe: A review of problems and opportunities for crop production and the environment. Soil Till. Res. **118**: 66-87. doi: http://dx.doi.Org/10.1016/j.still.2011.10.015.

Sörensen, R., Zinko, U., and Seibert, J. 2006. On the calculation of the topographic wetness index: evaluation of different methods based on field observations. J. Hydrol. Earth Syst. Sci. **10**(1):101-112. doi: https://doi.org/10.5194/hess-10-101-2006.

Sudduth, K., Kitchen, N., Wiebold, W., Batchelor, W., Bollero, G., Bullock, D., Clay, D., Palm, H., Pierce, F., Schuler, R.J.C., and Thelen, K.D. 2005. Relating apparent electrical conductivity to soil properties across the north-central USA. Comput. Electron. Agric. **46**(1-3): 263-283.   doi: https://doi.org/10.1016/j.compag.2004.11.010.

Sullivan, D.G., Shaw, J., and Rickman. 2005. IKONOS imagery to estimate surface soil property variability in two Alabama physiographies. Soil Sci. Soc. Am. J. **69**(6): 1789-1798.   doi: https://doi.org/10.2136/sssaj2005.0071.

Sun, W.-X., Huang, B., Qu, M.-K., Tian, K., Yao, L.-P., Fu, M.-M., and Yin, L.-P. 2015. Effect of Farming Practices on the Variability of Phosphorus Status in Intensively Managed Soils. Pedosphere 25**: 438-449. doi: https://doi.org/10.1016/S1002-0160(15)30011-4.

Tou, J.T., Gonzalez, R.C., and Gonzalez, R.C. 1974. Pattern Recognition Principles, Applied Mathematics and Computation.

Wang, C., Nolin, M., and Wu, J. 1995. Microrelief and spatial variability of some selected soil properties on an agricultural benchmark site in Quebec, Canada. Site-specific management for agricultural systems: Wiley Online Library. ASA-CSA-SSSA, Madison, WI, USA. pp. 339-350.

Whelan, B., and Mcbratney, A. 2000. The "null hypothesis" of precision agriculture management. Precis. Agric. **2**: 265-279. doi: 10.1023/A:1011838806489.

Williams, B.G., and Hoey, D. 1987. The use of electromagnetic induction to detect the spatial variability of the salt and clay contents of soils. Soil Res. **25**(1): 21-27. doi: https://doi.org/10.1071/SR9870021.

Wilson, H.F., Satchithanantham, S., Moulin, A.P., and Glenn, A. J. 2016. Soil phosphorus spatial variability due to landform, tillage, and input management: A case study of small watersheds in southwestern Manitoba. Geoderma **280**: 14-21. doi: http://dx.doi.Org/10.1016/j.geoderma.2016.06.009.

Wu, C., Wu, J., Luo, Y., Zhang, L., and Degloria, S.D. 2009. Spatial prediction of soil organic matter content using cokriging with remotely sensed data. Soil Sci. Soc. Am. J.   **73**(4): 1202-1208. doi: https://doi.org/10.2136/sssaj2008.0045.

Xin-Zhong, W., Guo-Shun, L., Hong-Chao, H., Zhen-Hai, W., Qing-Hua, L., Xu-Feng, L., Wei-Hong, H., and Yan-Tao, L. 2009. Determination of management zones for a tobacco field based on soil fertility. Comput. Electron. Agric. **65**(2): 168-175. doi: https://doi.org/10.1016/j.compag.2008.08.008.

Yao, R., Yang, J., Zhang, T., Gao, P., Wang, X., Hong, L., and Wang, M. 2014. Determination of site-specific management zones using soil physico-chemical properties and crop yields in coastal reclaimed farmland. Geoderma **232**: 381-393. doi: https://doi.org/10.1016/j.geoderma.2014.06.006.

Yates, S., and Warrick, A. 1987. Estimating soil water content using cokriging. Soil Sci. Soc. Am. J. 51**(1): 2330. doi: https://doi.org/10.2136/sssaj1987.03615995005100010005x.

Yuan, Y., Miao, Y., Yuan, F., Ata-Ui-Karim, S.T., Liu, X., Tian, Y., Zhu, Y., Cao, W., and Cao, Q. 2022.

Delineating soil nutrient management zones based on optimal sampling interval in medium-and smallscale intensive farming systems. Precis. Agric. 1-21. doi: 10.1007/s11119-021-09848-1.

Zebarth, B.J., Bélanger, G., Cambouris, A.N., and Ziadi, N. 2012. Nitrogen fertilization strategies in relation to potato tuber yield, quality, and crop N recovery. Sustainable potato production: global case studies : 165-186. doi: 10.1007/978-94-007-4104-1 10.

Ziadi, N., and Tran, T. 2008. Mehlich 3-Extractable Elements. Pages 81-88 in R. Carter and E. G. Gregorich, eds. Soil Sampling and Methods of Analysis. Taylor & Francis, Boca Raton, FL, USA.

4.10. Tables and Figures

Table 4-1. Mean air temperature and precipitation during the growing season (May-November) from 2012 to 2016 and 30-year normal (1981-2010) based on the St Hyacinthe 2 weather station (45°34.00 N, 72°55.00 W), Montérégie region.

	P_{rec}ipit ti$_{aon}$ (mm)					Ai_r temperature (C)				
	2012	2013	2015	2016	30 year Normal	2012	2013	2015	2016	30 years Normal
May	142.3	150.2	114.2	46	89.3	15.9	15.5	16.4	14.0	13.4
June	55.5	128.0	130.5	89.4	102.2	19.3	18.0	17.3	19.0	18.6
July	78.4	33.4	132.7	129.7	103.8	21.7	21.0	20.6	20.8	20.8
August	46.7	99.0	96.3	172.7	103.5	21.8	19.1	20.4	20.8	19.7
September	118.5	123.1	66.1	32.2	82.8	15.5	16.1	18.3	16.6	15.2
October	116.3	82.5	96.1	158.4	101.5	10.7	10.2	7.4	9.1	8.3
November	19.0	85.3	69.3	93.9	98.0	0.3		1.04.5	3.6	1.9
Total	576.7	701.5	705.2	722.3	681.1					
Average						15.0	14.4	15.0	14.8	14.0

Source: https://climat.meteo.gc.ca/climate normals/results 1981 2010 f.html?stnID=5492&autofwd=1 consulted on 14/02/2023.

Table 4-2. Descriptive statistics of the soil chemical properties for the soil layer (0-5 cm) and auxiliary properties

0-5 cm

	Unit	n	Mean	Min	Max	STD	CV[a] (%)
Soil chemical properties							
Soil pH$_{wa}$ ter		134	6.6	5.6	7.8	0.5	8
Total carbon	g kg ^{1}	134	23	15	35	3	15
P$_{M3}$	mg l<g ^{1}	134	80	21	195	29	36
Al$_{MS}$	mg l<g ^{1}	134	970	367	1089	94	10
Te$_{M3}$	mg l<g ^{1}	134	308	110	394	48	16
Ca$_{M3}$	mg l<g ^{1}	134	2218	1584	4569	469	21
(P/A1)$_{M3}$	(%)	134	7.9	2.5	17.5	2.6	32

	Depth (m)	Unit	n	Mean	Min	Max	STD	CV (%)
Soil elevation and TWI measured by Lidar								
Elevation		meters	134	34.8	34.2	35.0	0.1	0.3
TWI			134	10	7	14	1.1	11
Soil electrical conductivity measured by VERIS								
EC$_{a30}$	0-0.3	niSm ^{1}	2173	15.8	9.97	29.18	2.5	16
EC$_{a100}$	0-1.0	mS m ^{1}	2173	32.6	19.97	43.61	4.6	9
Crop yields measured by DGPS using yield monitor								
Yield com oi2$_2$		Mg ha ^{1}	6414	11.1	5.2	13.9	0.9	8
Yield soybean2013		Mg ha ^{1}	4642	3.6	2.3	4.6	0.3	9
Yield soybean2015		Mg ha ^{1}	4895	3.2	1.5	6.1	0.5	16
Yield com oi6$_2$		Mg ha ^{1}	6875	12.8	9.1	18.8	1.1	9

Note:[a] CV: Coefficient of variation. TWI: Topographical Wetness Index

P$_{M3}$; A1$_{M3}$; Fe | v | 3 and Ca$_{M3}$: available P, Al, Fe and Ca concentrations (mg kg$^{\wedge1}$) extracted using Mehlich-3 solution. (P/A1)$_{M3}$: ratio (%) determined from soil P$_{M3}$ and A1$_{M3}$ concentrations.

EC$_{a30}$; EC$_{a100}$: Electrical conductivity (mS nт^1) measured using Veris at 0-30 cm and 0-100 cm, respectively.

Table 4-3. Geostatistical parameters of the soil chemical and auxiliary properties.

0-5 cm

Model	Sill ratio[b] (%)	Range[c] (m)	R cv[2 d]	
Soil properties				
Soil pH_{water}	Sph	86	57	0.29
Total carbon	Sph	45	61	0.07
P_{M3}	Sph	43	55	0.06
Al_{M3}	Sph	52	64	0.27
Fe_{M3}	Sph	71	69	0.30
Ca_{M3}	Sph	99	77	0.26
$(P/Al)_{M3}$	Sph	46	60	0.08
Soil elevation and TWI measured by Lidar (n = 2173)				
Elevation	Sph	88	98	0.97
Topographical Wetness Index	Sph	78	242	0.50
Soil apparent electrical conductivity (ECa) measured by Veris (n = 2173)				
$ECa30$	Exp	78	40	0.84
$ECa100$	Exp	99	42	0.77
Crop yields measured by DGPS using yield monitors				
Yield $corn2012$	Exp	99	11	0.33
Yield $soybean2013$	Exp	58	16	0.38
Yield $soybean2015$	Exp	77	47	0.56
Yield $corn2016$	Sph	96	9.4	0.25

Note:[a] Semivariogram model: Sph, spherical, Exp, Exponential

[b] Sill ratio (%) = $[C/(c_0+c)]$ x100; this ratio measures spatial dependence or structure according to Whelan and McBratney (2000).

[c] Distance at which a semivariance becomes constant.

[d] Coefficient of determination of cross-validation.

P_{M3}; Al_{M3}; Fe_{M3} and Ca_{M3}: available P, Al, Fe and Ca concentrations (mg kg^{-1}) extracted using Mehlich-3 solution. $(P/Al)_{M3}$: ratio (%) determined from soil P_{M3} and Al_{M3} concentrations.

TWI: Topographical Wetness Index

$ECa30$; $ECa100$: Electrical conductivity (mS m^{-1}) measured using Veris at 0-30 cm and 0-100 cm, respectively.

DGPS: Differential global positioning system receiver

Table 4-4. Multiple spatial regression equations calculated for soil agri-environmental P index $(P/Al)_{M3}$ in relationship with soil chemical and auxiliary properties in the 0-5 cm soil layer

	Spatial regression equations	0-5 cm	R^2
Soil chemical properties			
Soil pH_{water}	$(P/Al)_{M3}= 7.1\text{x}10^{-2}\,pH_{water} +1.59$	$(p<0.01)$	0.210^{**}
Total carbon	$(P/Al)_{M3}= 2.6\text{x}10^{-2}\,TC+1.48$	$(p<0.01)$	0.130^{**}
Fe_{M3}	$(P/Al)_{M3}= -4.7\text{x}10^{-3}\,Fe_{M3}+9.43$	$(p=0.309)$	0.052
Ca_{M3}	$(P/Al)_{M3}= 8.7\text{x}10^{-4}\,Ca_{M3}+6.01$	$(p<0.0001)$	0.350^{***}
Auxiliary properties			
$ECa30$	$(P/Al)_{M3}= 9.1\text{x}10^{-5}\,ECa30+2.06$	$(p=0.978)$	0.047
$ECa100$	$(P/Al)_{M3}= 1.5\text{x}10^{-3}\,ECa100+2.11$	$(p=0.407)$	0.017
Yield $corn2012$	$(P/Al)_{M3}= -7.4\text{x}10^{-2}\,yield2012+8.77$	$(p=0.483)$	0.006
Yield $soybean2013$	$(P/Al)_{M3}= 0.352\,yield2013+6.69$	$(p=0.280)$	0.003
Yield $soybean2015$	$(P/Al)_{M3}= 0.306\,yield2015+6.93$	$(p<0.05)$	0.008^{*}
Yield $corn2016$	$(P/Al)_{M3}= 0.033\,yield2016+7.54$	$(p=0.744)$	0.029

Note: P_{M3}; Al_{M3}; Fe_{M3} and Ca_{M3}: available P, Al, Fe and Ca concentrations (mg kg^{-1}) extracted using Mehlich-3 solution. $(P/Al)_{M3}$: ratio (%) determined from soil P_{M3} and Al_{M3} concentrations.

TC: Total carbon content measured using an Elementar VarioMAX CN analyzer (TC, g kg^{-1}).

$ECa30$; $ECa100$: Electrical conductivity (mS m^{-1}) measured using Veris at 0-30 cm and 0-100 cm, respectively.

Yields: Crop yields during the 2012, 2013, 2015 and 2016 years.

$*$, $**$, $***$ are equivalent to p-value <0.05, $p < 0.01$, and $p < 0.001$, respectively.

Table 4-5. Mean values of soil chemical (0-5 cm) and auxiliary properties among two or three delineated MZs using three delineation strategies under the no-tillage field.

Mean values of soil chemical and auxiliary properties at (0-5 cm) depth

Number of MZ	Delineation strategies	MZs	$(P/A1)_{M3}$ (%)	$ECaso$ $mS\ m^{-1}$	TC $g\ kg^{-1}$	P_{M3} $mg\ kg^{-1}$	F_{CM3} $mg\ kg^{-1}$	pHwater	Ca_{M3} $mg\ kg^{-1}$	Yield2012 $Mg\ ha^{-1}$	Yield2013 $Mg\ ha^{-1}$	Yield2015 $Mg\ ha^{-1}$	Yield2016 $Mg\ ha^{-1}$
2MZ	$(P/A1)_{M3}$	MZ_1 6.6 b		15.6 a	2.2 b	68 b	314 a	6.4 b	2075 b	11.1 a	3.6 a	3.3 a	12.7 a
		MZ_2 10.1 a		16.1 a	2.4 a	100 a	300 b	6.8 a	2471 a	11.1 a	3.6 a	3.1 b	12.9 a
	ECa_{30}	MZ_i 7.8 b		14.4 b	2.3 a	75 b	319 a	6.5 b	2144 b	11.2 a	3.6 a	3.2 b	13.0 a
		MZ_2 8.4 a		18.1 a	2.3 a	81 a	306 b	6.7 a	2356 a	11.0 a	3.6 a	3.3 a	12.6 b
	$(P/Al)M$ $+ECa_{330}$	MZ_i 7.6 b		14.4 b	2.2 a	74 b	320 a	6.5 b	2102 b	11.2 a	3.6 a	3.2 a	12.9 a
		MZ_2 8.6 a		17.9 a	2.3 a	82 a	306 b	6.8 a	2405 a	11.0 a	3.6 a	3.3 a	12.6 b
3MZ	$(P/A1)_{M3}$	MZ_1 6.5 c		15.5 a	2.2 b	67 c	313 a	6.4 b	2073 b	11.1 a	3.6 a	3.3 a	12.7 a
		MZ_2 8.5 b		15.7 a	2.3 ab	87 b	319 a	6.6 b	2242 b	11.1 a	3.6 a	3.1 b	12.9 a
		MZ_3 11.8 a		16.6 a	2.4 a	111 a	283 b	7.0 a	2717 a	11.0 a	3.6 a	3.2 ab	12.9 a
	ECa_{30}	MZ_i 7.5 b		13.5 c	2.2 a	72 b	319 a	6.5 b	2072 c	11.2 a	3.6 a	3.2 ab	13.0 a
		MZ_2 8.1 a		15.8 b	2.3 a	78 a	315 a	6.6 ab	2235 b	11.1 a	3.6 a	3.1 b	12.9 a
		MZ_3 8.5 a		19.2 a	2.3 a	82 a	303 b	6.8 a	2426 a	11.0 a	3.6 a	3.4 a	12.5 b
	$(P/Al)M$ $+ECa_{330}$	MZ_i 7.2 b		13.9 c	2.3 b	70 b	319 a	6.5 b	2037 b	11.2 a	3.6 a	3.2 b	12.9 a
		MZ_2 8.5 a		15.9 b	2.3 a	82 a	317 a	6.6 a	2306 a	11.1 a	3.6 a	3.1 b	12.9 a
		MZ_3 8.5 a		19.1 a	2.3 a	81 a	299 b	6.8 a	2406 a	11.0 a	3.6 a	3.4 a	12.5 b

Note: $(P/Al)_{M3}$: Agri-environmental indicator ratio (%) calculated from soil P_{M3} and $A1_{M3}$ concentrations; ECa_{30}: Electrical conductivity measured at 0-30 cm ($mS\ m^{-1}$); TC: Total carbon content measured using an Elementar VarioMAX CN analyzer (TC, $g\ kg^{-1}$); P_{M3}; Fe_{M3} and Ca_{M3}: available P, Fe and Ca concentrations ($mg\ kg^{-1}$) extracted using Mehlich-3 solution. Yields: Crop yields during the 2012, 2013, 2015 and 2016 years.

Table 4-6. Agri-environmental P gain (kg P2O5) calculated for a corn-soybean rotation based on $(P/A1)_{M3}$ value from each MZ delineation strategy compared to the mean value $(P/A1)_{M3}$ from the conventional method under the no-tillage field.

	Conventional method	strategy 1 $(P/A1)_{M3}$ measurements		strategy 2 ECa_{3} o measurements		strategy 3 $(P/Al)M$ $+ECa_{330}$	
Management Zone (MZ)	IMZ	2MZ $(MZ\ MZ\ )_{b2}$	3MZ $(MZ\ MZ\ MZ\ )_{b213}$	2MZ $(MZ_b\ MZ\ MZ\ )_2$	3MZ $(MZ\ MZ_{b2i}\ MZ\ )_3$	2MZ $(MZ\ MZ\ )_{b2}$	3MZ $(MZ\ MZ_{b2i}\ MZ\ )_3$
Area[a] (MZ, ha)	MZ=9.5	MZ_1=5.77 MZ_2 =3.73	MZ_1=4.26 MZ_2 =3.31	MZ_1=6.0 MZ_2 =3.5	MZ_1=3.61 MZ_2	MZ_1=5.86 MZ_2	MZ_1=3.70 MZ_2

			MZ₃ = 1.93		=3.64 MZ₃ =2.25	=3.64	=3.57 MZ₃ =2.23
(P/A1)M3 value (%)	(P/A1)M3=7.9	(P/A1)M3(D=6.6%) (P/A1)M3($_2$)=10.1%	(P/A1)M3(D=6.5% (P/A1)M3($_2$)=8.5% (P/A1)M3($_3$)=11.8%	(P/A1)M$_3$ (D=7.8%) (P/A1)M ($_{32}$)=8.4%	(P/A1)M$_3$ (D=7.5%) (P/A1)M ($_{32}$)=8.1%(P/A1)M ($_{33}$)=8.5	(P/A1)M$_3$ (D=7.6%) (P/A1)M ($_{32}$)=8.6%	(P/A1)M$_3$ (D=7.2%) (P/A1)M ($_{32}$)=8.5 (P/A1)M ($_{33}$)=8.5
P recommendations for[b] com (kgP O$_{25}$ ha)[1]	40	$_z$ι∩ 20	40 40 20	$_z$ιn 4U	40 40 40	$_z$ιn 4U	40 40 40
P recommendations for[c] com from each MZ (kgP2O5)	TQ∩ J Ov	231 75	170 132 on	240	144 146 O∩	234	148 143 80
Total P recommendations for com (kgp o)$_{25}$	380	306	341	380	380	380	380

Note: MZ: Management zones. (P/A1)м3: Agri-environmental indicator ratio (%) calculated from soil Pм3 and A1м3 concentrations; ECa30: Electrical conductivity (mS rтr[1]) measured using Veris at 0-30 cm.

[a]Area: specific area (ha) from each specific MZ measured using ArcGIS.

[b]P specific recommendation (kg P2O5 ha^1) corresponding to each MZ value of (P/AI)M3 from each specific area. This was recommended following to the *Guide de Référence en Fertilisation du Québec* (CRAAQ, 2010). These P specific recommendation values were 0 kg P2O5 for soybean.

[c]P accurate recommendation (kg P2O5) applied in each MZ area (A), based on each MZ value of (P/AI)M3.

[d]Total P recommendation (kg P2O5) applied from all specific MZ areas: MZ₁, MZ2 and MZ3

98

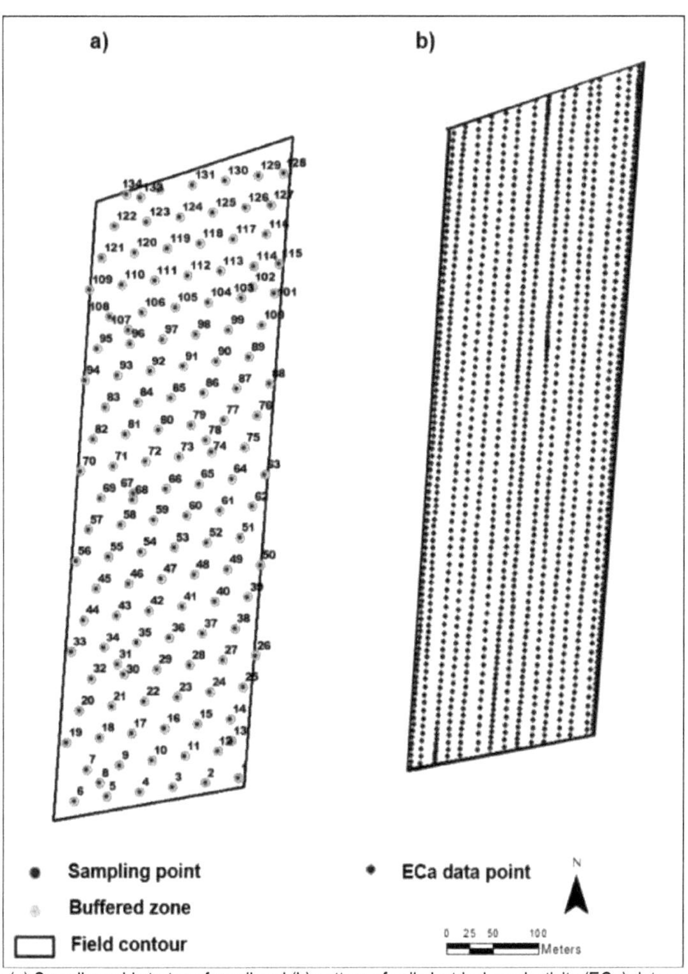

Figure 4-1: (a) Sampling grid strategy for soil and (b) pattern of soil electrical conductivity (ECa) data acquisition.

Figure 4-2. Flowchart summarizing the delineation strategies using ISODATA method. **Note:** MZ: Management zones; *nMZ*: optimal number of MZs. $_{PM3}$ and $_{AIM3}$: soil available P and Al concentrations (mg kg^{-1}) extracted using Mehlich-3 solution; (P/Al)$_{M3}$: ratio (%) determined by soil $_{PM3}$ and $_{AIM3}$ concentrations. ECa: Electrical conductivity (mS m^{-1}) measured using Veris. ANOVA: Analysis of Variance determined using SAS program. P2O5 fertilizer recommendations: P fertilizer recommendation (kg P2O5 ha^{-1}) applied from all specific MZ areas resulting from each delineation strategy

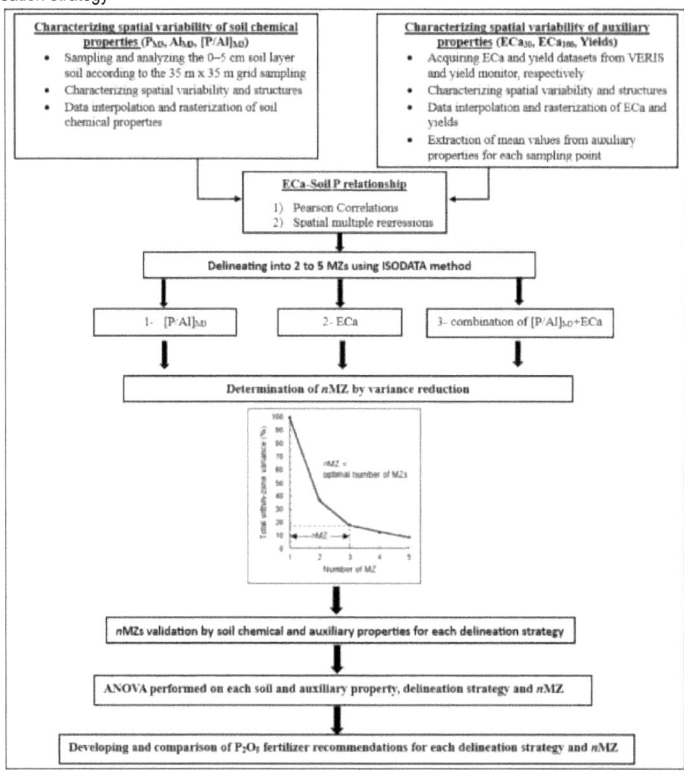

Figure 4-3. Spatial distribution maps of soil chemical properties PM3 (a), (P/A1)M3 (b), A1M3 (c), TC (d), PeM3 (e) and CaM3 (f) for the 0-5 cm soil layer. Note: PM3; A1M3; Fe | v | 3 and CaM3: available P, Al, Fe and Ca concentrations (mg kg^1) extracted using Mehlich-3 solution. (P/A1)M3: ratio (%) determined from soil PM3 and A1M3 concentrations. TC: Total carbon content measured using an ElementarVarioMAXCN analyzer (TC, g kg1).

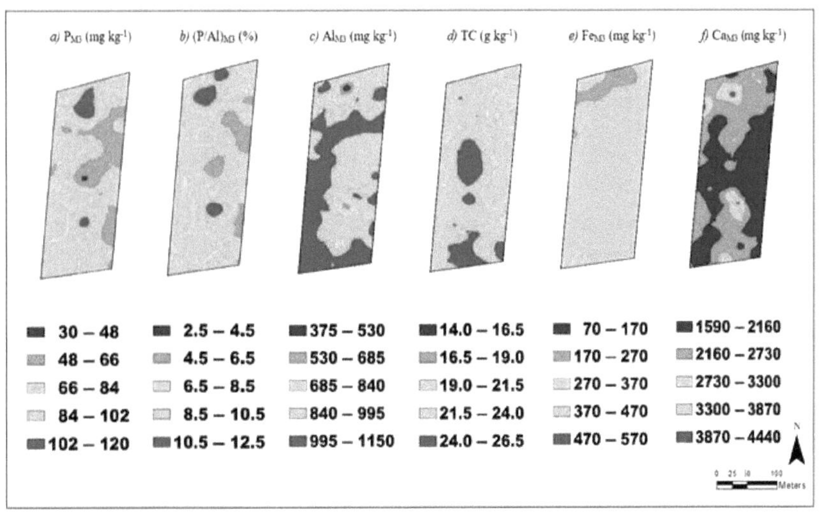

Figure 4-4. Spatial distribution maps of auxiliary properties ECa30 (a), ECawo (b), Yield corn2012 (c), Yield soybean2013 (d), Yield soybean20w (e) and Yield corn20i6 (f). Note: ECa30; ECawo: Electrical conductivity (mS nT1) measured using Veris at 0-30 cm and 0-100 cm, respectively. Yields: Crop yields during the 2012, 2013, 2015 and 2016 years.

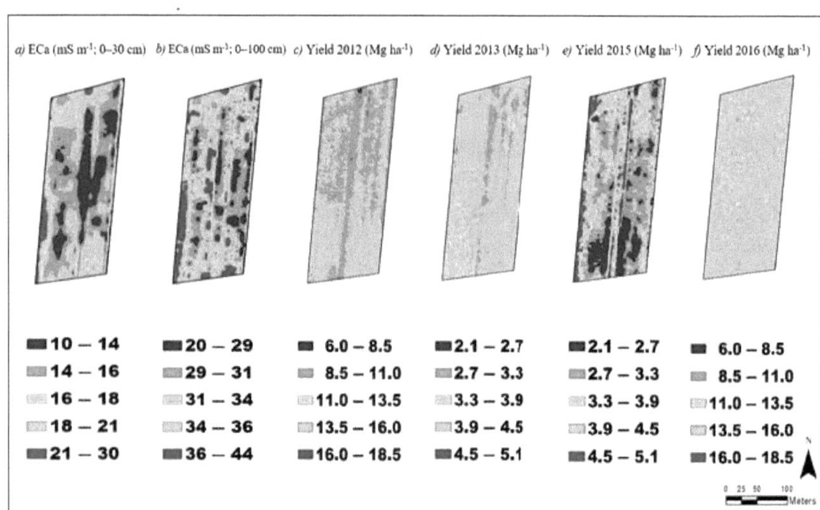a) ECa (mS m⁻¹; 0–30 cm)   b) ECa (mS m⁻¹; 0–100 cm)   c) Yield 2012 (Mg ha⁻¹)   d) Yield 2013 (Mg ha⁻¹)   e) Yield 2015 (Mg ha⁻¹)   f) Yield 2016 (Mg ha⁻¹)
10 — 14
14 — 16
16 — 18
18 — 21
21 — 30
20 — 29
29 — 31
31 — 34
34 — 36
36 — 44
6.0 – 8.5
8.5 – 11.0
11.0 – 13.5
13.5 – 16.0
16.0 – 18.5
2.1 – 2.7
2.7 – 3.3
3.3 – 3.9
3.9 – 4.5
4.5 – 5.1
2.1 – 2.7
2.7 – 3.3
3.3 – 3.9
3.9 – 4.5
4.5 – 5.1
6.0 – 8.5
8.5 – 11.0
11.0 – 13.5
13.5 – 16.0
16.0 – 18.5
0 25 50 100
Meters

Figure 4-5. Pearson correlation coefficients (r) of soil agri-environmental P index (P/Al)M3 and apparent electrical conductivity (ECa30 and ECa100) in relationship with soil chemical and auxiliary properties for the 0-5 cm soil layer. **Note**: PM3; AlM3; FeM3 and CaM3: available P, Al, Fe and Ca concentrations (mg kg^{-1}) extracted using Mehlich-3 solution. (P/Al)M3: ratio (%) determined from soil PM3 and AlM3 concentrations. ECa30; ECa100: Electrical conductivity (mS m^{-1}) measured using Veris at 0-30 cm and 0-100 cm, respectively. Yields: Crop yields during the 2012, 2013, 2015 and 2016 years. *, **, *** are equivalent to p-value <0.05, p <0.01, and p <0.001, respectively. .

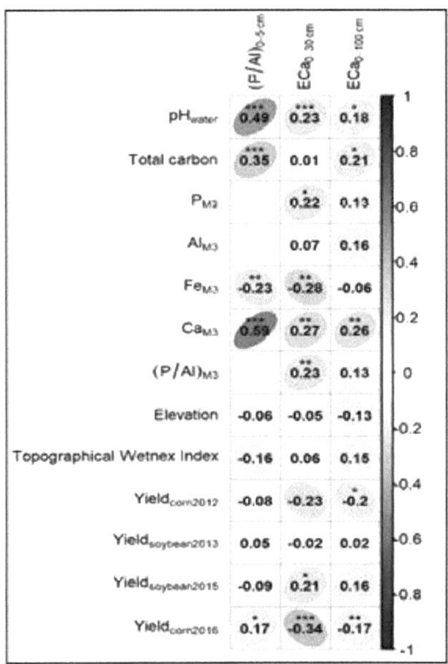

Figure 4-6. Delineation of MZs (2-5 MZ) using (P/A1)M3 measurements (a-d), ECa30 measurements (e-h) and (P/Al)M3+ ECa30 (i-l) strategies in the 0-5 cm soil layer. **Note:** MZ: Management zones; (P/A1)M3: ratio (%) determined from soil PM3 and A1M3 concentrations. ECa30: Electrical conductivity (mS n⊤$^{-1}$) measured using Veris at 0-30 cm

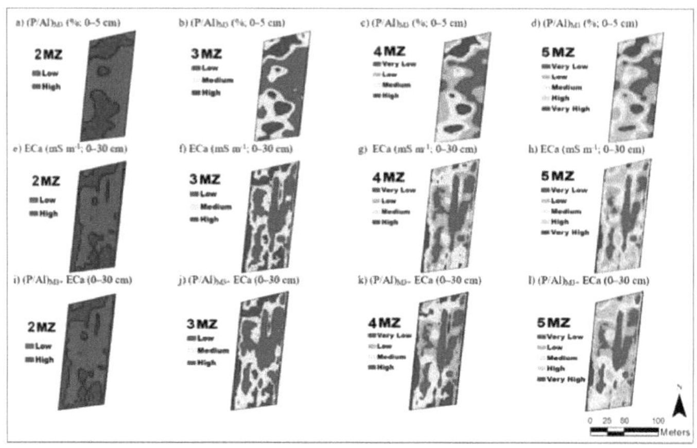

Figure 4-7. Reduction of total within-zone variance for (P/Al)M3, soil apparent electrical conductivity (ECa30), soil chemical properties (0-5 cm) and crop yields (2012-2016) into MZ using (P/Al)M3 (a-d) and ECa (0-30 cm) measurements (e-h) delineation strategies. **Note:** MZ: Management zones. pH: soil pH measured using a 1:1 soil-water ratio. PM3; FeM3 and CaM3: available P, Fe and Ca concentrations (mg kg⁻¹) extracted using Mehlich-3 solution. (P/A1)M3: ratio (%) determined from soil PM3 and A1M3 concentrations. TC: Total carbon content measured using an Elementar VarioMAX CN analyzer (TC, g kg⁻¹). ECa30: Electrical conductivity (mS m^1) measured using Veris at 0-30 cm. Yields: Crop yields during the 2012, 2013, 2015 and 2016 years.

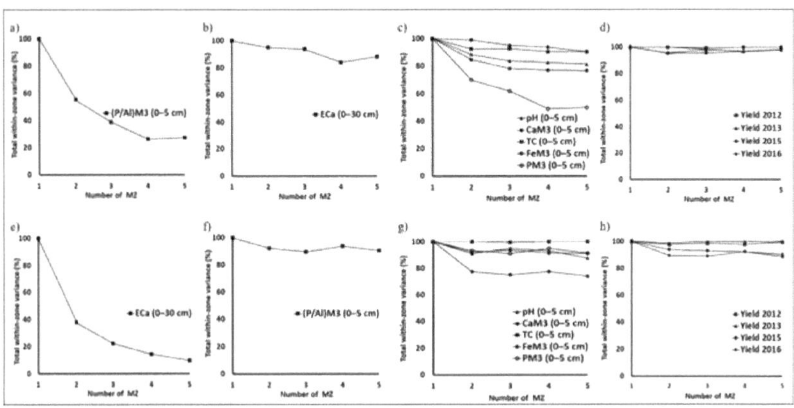

Figure 4-8. Reduction of total within-zone variance for (P/A1)м3 and soil apparent electrical conductivity (ECa30) (a), soil chemical properties (0-5 cm) (b) and crop yields (2012-2016) (c) into MZ using (P/Al)м3+ ECa30 delineation strategy. **Note:** MZ: Management zones. pH: soil pH measured using a 1:1 soil-water ratio. Pм3; Feм3 and Caм3: available P, Fe and Ca concentrations (mg kg^{-1}) extracted using Mehlich-3 solution. (P/A1)м3: ratio (%) determined from soil Pм3 and A1м3 concentrations. TC: Total carbon content measured using an Elementar VarioMAX CN analyzer (TC, g kg^{-1}). ECa30: Electrical conductivity (mS пт^{1}) measured using Veris at 0-30 cm. Yields: Crop yields during the 2012, 2013, 2015 and 2016 years.

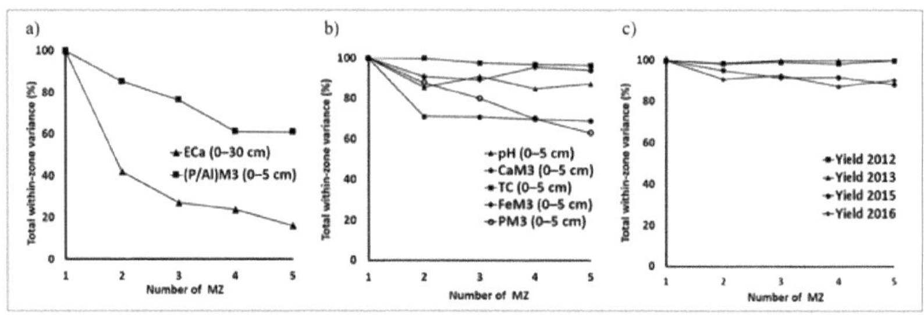
a)
Total within-zone variance (%)
ECa (0–30 cm)
(P/Al)M3 (0–5 cm)
Number of MZ
b)
Total within-zone variance (%)
pH (0–5 cm)
CaM3 (0–5 cm)
TC (0–5 cm)
FeM3 (0–5 cm)
PM3 (0–5 cm)
Number of MZ
c)
Total within-zone variance (%)
Yield 2012
Yield 2013
Yield 2015
Yield 2016
Number of MZ

Chapter 5

General discussion

P is an important nutrient for crop growth and development. However, excessive applications of P to agricultural soils in relation to crop requirements have resulted in stratification of P, increasing the risk of loss of this element through surface runoff (Jamieson et al., 2003). This contributes to environmental degradation through eutrophication of watercourses (Grant et al., 1996; Jordan et al., 2000; Messiga, 2010; Wei-Xia et al., 2015).

Thus, sustainable P management in arable soils relies on fertilisation that balances the soil's actual P supply and demand (Tunney, 1990; Fu et al., 2013a). It is therefore important to have a better understanding of the spatial variability (SV) of P in arable soils, in order to improve the economic and rational use of P fertilizers, promote the profitability and sustainability of the farming enterprise, improve the reduction of P losses and the preservation of the environment. To date, there are few studies on the impact of cropping systems and tillage on P SV at the farm field level in the province of Quebec. In addition, few studies have been carried out on the delimitation of P-based management zones (ZAs) using apparent soil electrical conductivity (ECa) for the purpose of reducing P SV in Quebec.

In this thesis, the impacts of cropping systems and tillage on P SV were assessed for the purpose of precise P_2O_5 recommendations to reduce P losses. In addition, the delimitation of P-based ZAs was assessed using ECa to better reduce this variability in a precision agriculture context. Thus, four main aspects will be summarised in this general discussion, namely 1- P stratification and its agri-environmental implications; 2- agronomic factors affecting P SV; 3- relationships between P and other physico-chemical and auxiliary properties; and 4- agri-environmental benefits associated with better P management at the field scale.

5.1 Stratification of P and its agri-environmental implications

P stratification remains a major agri-environmental issue for soils in the Province of Quebec. The results in Chapter 2 showed that a high application rate of P-rich organic amendments resulted in P stratification under permanent grassland (10 years), compared with young grassland (2 years), where P concentrations were similar in both soil layers (0-5 cm and 520 cm) (Table 2-1). Moreover, spatial P values were higher in the 0-5 cm layer than in the 520 cm layer of the soil under PA, while they were similar in the soil under JP (Figure 2-2; Figure 2-3), confirming the stratification of P. Permanent grassland (10 years) increased the agri-environmental risk of P pollution compared with soil under young grassland (2 years). Spatial distribution maps of P indicators (P_{M3} and $(P/Al)_{M3}$) showed that high $(P/Al)_{M3}$ levels in the soil under permanent grassland were close to the agri-environmental risk of P pollution (threshold limit=15%) (Figure 2-3) compared with the average $(P/Al)_{M3}$ value (7%) in the field (Table 2-1) generally considered for P_2O_5 recommendations in the province of Quebec.

Similarly, the results in Chapter 3 also showed that the agronomic practice of no-till (NTM) resulted in stratification of P, compared with conventional tillage (CT) where P concentrations were similar in both soil layers (0-5 cm and 5-20 cm) (Figure 3-2). In addition, spatial P values were higher in the 0-5 cm layer than in the 5-20 cm layer of the soil under SD, whereas they were similar in the soil under CT (Figure 3-3; Figure 3-4), confirming the stratification of P. Thus, the agronomic practice of SD increased the agri-environmental risk of P pollution (8%) compared with soil under CT. Spatial P values revealed that agri-environmental soil P indicator levels under SD were higher than the agri-environmental limit threshold for P pollution (threshold=8%) (Figure 3-4) compared with the average field $(P/Al)_{M3}$ value (8%) (Figure 3-2) often considered for P_2O_5 recommendations in the province of Quebec.

Thus, Chapters 2 and 3 demonstrated the importance of P stratification and its agri-environmental implications, in terms of P pollution risks under the impact of different cropping systems and tillage. Many agri-environmental concerns have been observed regarding SD practice in relation to P stratification (Cade-Menun et al., 2010; Messiga et al., 2010; Messiga et al., 2012; Abdi et al., 2014), transport and runoff to surface waters (Puustinen et al., 2005; Allaire et al., 2011), which can lead to P eutrophication in rivers (Jordan et al., 2000; Sun et al., 2015; Patoine et al., 2017).

These two chapters also revealed the development of spatial P heterogeneity according to the soil P distribution model at field scale, as confirmed by previously conducted studies (Fu et al., 2013a; Peralta and Costa, 2013;

Bogunovic et al., 2014). This spatial P heterogeneity results from high soil P variability, hence the importance of considering P SV and agronomic factors that may affect this variability for accurate P2O5 recommendations for P loss reduction purposes.

5.2 Agronomic factors affecting spatial variability of P at field scale

Farming systems

The results in Chapter 2 showed that, based on CV, P variability was high to very high (45-81%) according to the classification of Nolin and Caillier (1992b) in relation to other physical chemical soil properties (Table 2-1). West et al (1989) explained that high P variability under grassland is probably due to the presence of highly labile chemical forms of P in the soil. These CV values obtained are consistent with previous studies carried out under grassland in Ireland (McCormick et al., 2009; Fu et al., 2013a; Fu et al., 2013b).

The results of Chapter 2 also revealed that the permanent grassland system (10 years) combined with intensive fertilisation practices modified P variability at field level (Table 21). P variability was higher (63-81%) under JP (2 years) in a maize-soybean rotation, while it was lower (46-64%) under permanent grassland. Thus, permanent grassland (10 years) reduced P variability at field level. The study by Jia et al (2011) also observed the same effects.

Furthermore, the high variability of P due to the effect of grassland systems combined with intensive fertilisation practices has an impact on two agronomic implications, namely (1) the sampling strategy based on the number of samples required (n) and (2) the non-uniformity of P recommendations (kg P2O5). High P variability corresponds to a high number (n) of samples required (sampling density) in order to achieve a given accuracy in estimating the mean P (Wilding and Drees, 1983; Nolin et al., 1991), according to the following formula: $n=t^2$ CV /ER[22] . n is a function of the variability (CV) of P, the relative error (RE) and the probability chosen. In this way, the relationship between sampling, intensity and the probability that the field average of a given sample size remains within an acceptable range (Nolin et al., 1991; Grandt et al., 2010). The aim is to better capture this variability.

The high variability of P means that P2O5 recommendations are not uniform, which is why it is important to take this into account in order to apply the right amount of P, at the right time, in the right place and in the right way in order to reduce P losses. P variability based on CV is a good indicator of the intensity of P variation, but it does not tell us anything about the nature of this variability (structured or randomised) or the precise pattern of this variability (Cambouris et al., 2006).

The results in Chapter 2 showed that the spatial dependence of soil P under JP and AP was moderate (31-71%) (Table 2-2). This results from the interaction between intrinsic (soil series) and extrinsic soil factors (intensive P fertilisation or farming practices) that determine the spatial structure of P. These results obtained were in line with other studies previously carried out (Whelan and McBratney, 2000; Jia et al., 2011; Fu et al., 2013a).

The results of Chapter 2 also revealed that intensive fertilisation practices (P-rich amendments) reduced the spatial dependence of soil P under permanent grassland over a longer period (10 years) (Table 2-2). This reduced spatial dependence of P is explained by the increasing influence of extrinsic soil factors (P-rich fertilisation), which modified the long-term spatial structure of soil P in the field. Several studies have confirmed these results (Grandt et al., 2010; Jia et al., 2011; Fu et al., 2013a). A reduced spatial structure of soil P could have an impact on the efficiency of delimiting ZAs for sustainable soil P management in grasslands. Indeed, a high spatial structure of P will have a positive impact in terms of ZAs delimition, while a low spatial structure of P will make it difficult to use ZAs due to the lack of spatial dependence of P.

Working the soil

The results in Chapter 3 showed that, based on CV, P variability was moderate to very high (32-60%) (Nolin and Caillier, 1992b) compared with other physico-chemical soil properties (Figure 3-2). The variability of P was highest compared with other soil properties. These results were similar with other studies previously conducted under tillage (Cambouris et al., 2017; Dalchiavon et al., 2017; Malvezi et al., 2019). The high variability of P under direct seeding (DS) would be explained by (1) the reduced mixing of soil, crop residues and fertilizer P, and (2) the high levels of residual P at the soil surface (Kitchen et al., 1990; Tyler and Howard, 1991; Mallarino, 1996).

The results in Chapter 3 also revealed that the agronomic practice of SD reduced P variability at the field scale (Figure 3-2). The results are similar to other studies conducted at the scale of experimental plots in Quebec

(Cambouris et al., 2017). Reduced variability under SD would have a positive impact on the sampling strategy, characterised by a reduced number of soil samples.

The results in Chapter 3 showed that the spatial dependence of P in tilled soils was moderate (23-46%) (Table 3-2). This is a result of the interaction between intrinsic (soil series) and extrinsic soil (tillage) factors that determine the spatial structure of P. These obtained results were in line with other previously conducted studies (Dalchiavon et al., 2017). The spatial dependence of soil P was higher under SD compared to CT, due to reduced tillage under SD, impacting on the long-term spatial structure of soil P at the field scale.

Pedodiversity and sampling density

The work in this thesis also highlights the impact of low soil diversity on the variability and spatial dependence of P under the effect of cropping systems, tillage combined with intensive fertilisation practices. Previous studies on P SV were carried out in fields containing several soil series in eastern Canada (Nolin et al., 1999; Nolin et al., 2003; Cambouris et al., 2006; Perron et al., 2018). These studies were characterised by high P variability associated with high spatial dependence (r > 75%), resulting from high soil diversity (Cambouris et al., 2006; McCormick et al., 2009; Roger et al., 2014; Sun et al., 2015). The presence of long, narrow plots made up of one or more rounded beds explains much of the high variability of many soil properties, including P in the region (Quenum et al., 2012).

Thus, the results of Chapters 2 and 3 showed high P variability under the effect of cropping systems, tillage and low soil diversity at the field scale in eastern Canada (Table 2-1; Figure 3-2). Nevertheless, this high P variability is associated with reduced spatial P dependence (23-71%) compared to previous studies conducted under high soil diversity (Nolin et al., 1999; Nolin et al., 2003; Cambouris et al., 2006; Perron et al., 2018). This reduced spatial dependence on P is caused by P-rich fertilisation (Grandt et al., 2010), combined with low soil diversity. Furthermore, the work also demonstrated a reduction in the variability and spatial dependence of A_{LM3} (Table 2-1; Table 2-2; Figure 3-2; Table 3.2) compared with previous studies carried out, characterised by higher variability and spatial dependences of $1'A_{1M3}$ under high pedodiversity (Nolin et al., 2003; Perron et al., 2018).

Sampling density is a non-negligible factor in assessing the SV of P at field scale. A reduced sampling density has little effect on P variability (Radocaj et al., 2021), but has a strong impact on the spatial structure of P (Fu et al., 2013b). Shi et al (2000) obtained spatial structures of up to 98% corresponding to lower sampling densities (i.e. 2 samples ha^1). These results were also confirmed by Nze Memiaghe et al (2022). Furthermore, the results in Chapter 4 demonstrated that the ZAs delimitation method based on the sampling density of the $(P/A1)_{M3}$ represented the best ZAs delimitation strategy for the purposes of reducing (45%) the SV of the $(P/A1)_{M3}$ (Figure 4-6). Thus, sampling density remains a dominant factor in characterising the SV of P.

ECa: a limited tool for characterising the spatial variability of P

ECa represents the best modern technology for delimiting ZAs in the field (Cambouris et al., 2006). Indeed, it can be used as an auxiliary property to better characterise the intra-field SV of soil properties (Cambouris et al., 2006), due to its strong correlation with several physico-chemical soil properties (Sudduth et al., 2005; Saifuzzaman et al., 2021; Becker et al., 2022).

Nevertheless, the results in Chapter 4 demonstrated that ECa represented a limited tool for characterising P SV in this study, due to a low significant correlation (0.22-0.23) observed between ECa_{30} and the two P indicators in P-stratified soil (Figure 4-4). These results were similar with previous studies carried out in P-stratified soils (Omonode and Vyn, 2006; Peralta and Costa, 2013). The low intensity of the CEa-P correlation could be explained by low water content (Rhoades et al., 1989; Omonode and Vyn, 2006) associated with low pedodiversity observed in soils (Adamchuk et al., 2015).

Thus, soil ECa-P relationships vary from field to field and are probably due to the water content and performance of field-specific ECa sensors, where the field is dominated by one or two main intrinsic factors, such as clay content or soil moisture (Corwin and Lesch, 2005; Brevik et al., 2006). These two intrinsic soil factors make the interpretation of ECa values very specific to the field studied. However, ECa measurements remain an important tool for assessing soil P fertility in precision agriculture, as they have the potential to identify areas in fields where soil types, nutrient levels and productivity differ greatly (Omonode and Vyn, 2006).

5.3 Relationship between P and other physico-chemical and ancillary properties

P and the physico-chemical properties of soils

The results of chapters 2 and 3 showed that agronomic practices (cropping systems and tillage) had an impact on the correlations between P and soil physicochemical properties. In fact, the results of chapter 2 showed that the differences in correlation between P and soil physicochemical properties varied according to the contrasting grassland systems (Table 2-3; Table 2-4). Permanent grassland resulted in an accumulation of P, significantly modifying the effects of 1'$A1_{M3}$, Fe_{M3} and $CaM3$ on stratified P, resulting from a decrease in pH. These results were confirmed by the visual associations between P and these chemical elements in the spatial distribution maps (Figures 2-3 d, e and f).

Neoformed P-Ca and P-Fe bonds were also observed, confirmed by other previous studies (Beauchemin et al., 2003; Kuo et al., 2005; Eriksson et al., 2016).

The results in Chapter 3 showed that the relationships between P and soil physico-chemical properties differed according to tillage practices. Indeed, SD led to an accumulation of P, inducing significant changes in the relationships between P and pH, total carbon and $CaM3$, resulting in a drop in pH (Table 3-3; Table 3-4). This drop in soil pH is due to acidification from nitrogen fertilisers combined with decomposition of crop residues (Limousin et al., 2007; Soane et al., 2012; Malvezi et al., 2019). These results were confirmed by the visual associations between P and these chemical elements in the spatial distribution maps (Figures 3-4 a, b and f).

P and auxiliary properties

The results of article 3 showed that the $CEa30_{-P}$ relationship was weak (r < 0.70) according to Heiniger et al. (2003) at field scale (Figure 4-4). These results have an agronomic impact on the delimitation of ZAs, precluding a geostatistical co-kriging approach aimed at improving the prediction of the spatial distribution of $(PZAI)_{MS}$ using $CEa30$ on the 0-5 cm layer in this study. Thus, three distinct ZAs delineation strategies based on 1) $(P/Al)_{M3}$; 2) $CEa30$; and 3) the combination $(P/Al)_{M3+CEa30}$ (Figures 4-5) were developed to reduce the SV of $(P/Al)_{M3}$, aiming to optimise the precise P_2O_5 recommendations under SD.

No significant $CEa100_{-P}$ correlation was observed, similar to Peralta and Costa (2013). This was mainly due to two reasons: (1) the equivalent conductances of inorganic P ions ($H2PO4^-$, $HPO4^{2-}$) were generally lower than those of other ionic species (Ca^{2+} and Mg^{2+}) in soils; and (2) fertilisation practices (strip application) and direct seeding impacted on this weak soil $CEa100_{-P}$ relationship (Jung et al., 2005; Motavalli et al., 2015).

In this thesis, the study of P-yield relationships over four successive years constitutes an originality in the Province of Quebec. The results showed that P-yield relationships varied from year to year, as observed in previous studies (Guedes Filho et al., 2010; Diacono et al., 2012; Ameer et al., 2022; Nyéki et al., 2022). This relationship depends on two dominant factors, namely soil properties and climatic conditions, which have an impact on yield variability (Jiang and Thelen, 2004; Singh et al., 2016; Nyéki et al., 2022). Soil properties and climatic conditions being able to vary each year due to climatic variations (Rodrigues and Corá, 2015).

P variability and physico-chemical and auxiliary properties

Multiple spatial regression equations were generated to assess the impact of soil physicochemical and auxiliary properties on SV of P. Thus, the results of paper 1 revealed that permanent grassland accumulated P, significantly affecting the impact of $CaM3$ and $FeM3$ on P variability (Table 2-4). Similarly, the results of paper 2 revealed that SD practice accumulated P, inducing significant changes of CT, $A1_{M3}$, $FeM3$ and $CaM3$, on P variability (Table 3-4).

The results of Paper 3 (Table 4-4) showed that 13-35% of the variations in $(P/A1)_{M3}$ were significantly explained by $CaM3$, $p\frac{1}{4}au$ and TC. However, these results also showed that auxiliary properties had a small impact (0.3 - 5%) on the variability of $(P/A1)_{M3}$ (Table 4-4). Omonode and Vyn (2006) obtained similar results confirming the low impact of ECa (< 7%) on P variability in a P stratified soil under a 50-year maize-soybean rotation.

Thus, the application of the spatial regression equations confirmed the strength and stability of the relationships between the P indicators (P_{M3}; $[P/A1]_{M3}$) and $CaM3$, $FeM3$, and TC. Thus, it would be wise to consider the variability of these physicochemical properties and their respective spatial structures, in order to better understand the variability of accumulated soil P under these different agronomic practices, as these physicochemical properties have a significant impact on soil P accumulation.

5.4 Agri-environmental implications of spatial P variability: P2O5 gains in grain maize crops

Soil P SV can be controlled by two main agronomic strategies in precision agriculture: (1) variable rate application (VRA) and (2) the use of management zones (ZAs). Both approaches aim to reduce environmental P losses. Thus, the results of Article 2 evaluated the gains in P O_{25} for the grain maize crop based on ATV for the purposes of specific P recommendations O_{25} compared with uniform application based on the average value of the agri-environmental indicator (P/Al)M3 of the entire field (Table 3-5). In the field under SD, the total amount of P2O5 based on the average value of the agri-environmental indicator (P/A1)M3 was 380 kg P2O5 compared with 364 kg P2O5 based on ATV using georeferenced maps for P2O5 recommendations in grain maize. This generated a reduction of 14 kg P2O5 in the field.

Furthermore, the results of Article 3 evaluated P2O5 input reductions for grain maize using three distinct delimitation strategies based on 2 and 3 ZAs for the purpose of specific P2O5 recommendations compared to the conventional method based on the average (P/A1)M3 value of the whole field (380 kg P2O5) similar to Article 2 (Table 4-6). These 3 ZAs delineation strategies are based on 1) (P/Al)M3; 2) CEa30; and 3) the combination of (P/Al)M3+CEa30 in order to reduce the SV of (P/Al)M3, aiming to optimize the precise P2O5 recommendations under SD.

Thus, 6 agronomic P2O5 recommendations based on 2 and 3 ZAs were determined for each delimitation strategy, compared with the conventional method (380 kg P2O5) (Table 4-6). The total amounts of P2O5 based on 2 and 3 ZAs according to (P/Al)M3 represented 306 and 341 kg P2O5 in the field, respectively. This generated respective reductions of 74 and 39 kg P2O5 compared to the conventional method (380 kg) in the field. However, the total amounts of P2O5 based on 2 and 3 ZAs according to the 2 other delimitation strategies (CEa30; (P/Al)M3+CEa30) represented 380 kg P2O5, which generated no savings in P (0 kg P2O5), compared to the conventional method (380 kg).

In summary, the ATV generated 366 kg P2O5, while the delimitation by 2 and 3 ZAs generated 306 and 341 kg P2O5, i.e. respective gains of 60 and 25 kg P2O5 compared to the ATV in grain maize (Table 35; Table 4-6). Consequently, the 2 and 3 ZAs delimitation strategy appeared to be more effective than the ATV in terms of P gain O_{25} in this field under P accumulation. This study has demonstrated the importance of taking into account the SV of P for specific P management purposes, to reduce P losses in field crop soils in a precision agriculture context.

Conclusions

P is an important nutrient for crop growth and development. However, excessive applications of P to agricultural soils in relation to crop needs have led to stratification of P, increasing the risk of loss of this element through surface runoff and contributing to environmental degradation through eutrophication of watercourses. Sustainable P management in arable soils therefore depends on fertilisation that balances the soil's actual P supply and demand. It is therefore important to have a better understanding of the SV of P in crop soils, in order to improve the economic and rational use of P fertilisers, promote the profitability and sustainability of the farming enterprise, improve the reduction of P losses and preserve the environment. The general objective of this doctoral thesis was to evaluate soil P SV under different cropping systems and tillage for the purposes of agri-environmental P_2O_5 recommendations in precision agriculture.

The results in Chapter 2 showed that repeated applications of organic amendments resulted in an accumulation of available P [P_{M3}, $(P/A1)_{M3}$] in the 0-5 cm layer under former grassland. This accumulation of P increased the agri-environmental risk of P pollution in permanent grassland. The reduced variability and spatial dependence of soil P was observed in both soil layers (0-5 cm and 5-20 cm) under old grassland due to the increasing influence of extrinsic soil factors, such as organic fertiliser applications over a long period. From an agronomic point of view, a reduced spatial structure of soil P may have an impact on the delimitation of ZAs for soil P management purposes in grassland. The relationships between soil available P indices and other physico-chemical properties differed between the contrasting grassland systems.

The results of Chapter 2 also demonstrated that soil P spatial distribution maps should be used to guide variable rate applications (VRAs) of P fertiliser in order to avoid under- and over-fertilisation of P in grassland. According to the results of this study, a soil sampling strategy focusing on the top five centimetres of soil should be recommended in permanent grasslands, due to the high risk of P stratification caused by high P fertilisation, as opposed to the current soil sampling strategy (0-17.5 cm) applied in the Province of Quebec, for the purpose of precise P_2O_5 recommendations.

The results in Chapter 3 showed that the soil P indices measured were similar in both layers (0-5 cm and 5-20 cm) under CT. However, under SD, these P parameters were higher in the 0-5 cm layer compared with the 5-20 cm layer, confirming the stratification of P. Thus, the agronomic practice of SD resulted in an accumulation of P [$(P/Al)_{M3}$] in the soil (7.9%) compared with CT (2.7%), increasing the risk of P pollution, as observed in the spatial kriging maps.

The results in Chapter 3 also showed that the relationships between soil available P indices and other physico-chemical soil properties differed under contrasting tillage practices. The variability of soil P indices in the two fields ranged from moderate to very high (32-60%), indicating that the uniformly applied P fertiliser recommendations currently in use are not suitable for fields under large areas (10 ha). These results have agri-environmental implications linked to the sustainable management of soil P. In fact, ATV using krigée maps generated a reduction of 14 kg P_2O_5 in the layer (0-5 cm) of the field under SD for grain maize cultivation. In this chapter, the ATV demonstrated the importance of applying phosphate fertilisers from the right source, at the right dose, in the right way and in the right place, in order to draw up accurate future P recommendations for reducing P losses to the environment.

The results in Chapter 4 showed that the mean value of $(P/Al)_{M3}$ was 7.9% under the stratified layer (0-5 cm) of soil P under SD. The mean values of CEa_{30} and CEa_{100} were 15.8 and 32.6 mS m^{-1} respectively. The CV of the soil P indices remained high (32-36%), despite low soil diversity in the field. A high intensity of P variation in a series of soils has an impact on the delineation of homogeneous P-based ZAs in a stratified P field, for specific P fertiliser management purposes. The spatial dependence of soil P and ancillary properties (CEa measurements and agronomic yields) varied from moderate to high (43-99%) as a result of the interaction between intrinsic and extrinsic factors (*i.e.* climatic variability).

Variable relationships between $(P/Al)_{M3}$, ECa measurements and yields (2012, 2015 and 2016) were observed due to inter-seasonal climatic variability (Figure 4-5). P-yield relationships varied from year to year due to two main dominant factors (soil properties and climatic conditions), impacting yield variability. Similarly, EC-yield relationships were observed, resulting from yield variability. A significant relative CE_{a30-P} soil correlation (r of 0.22-0.23) was observed, generating three ZAs delimitation strategies.

The results of Chapter 4 also showed that there is potential for reducing the spatial variability of soil P using the ZAs delimitation method, an agronomic strategy for controlling spatial variability, aimed at reducing P losses while maintaining yields in the Province of Quebec.

In accordance with the results of this study, delimiting the stratified P field into two or three ZAs using $(P/Al)_{M3}$ measurements represented the best agronomic strategy for controlling the spatial variability of $(PZAI)_{Ms}$. This was due to the greatest reduction (45%) in the spatial variability of $(PZAI)_{M3}$ (1-2 ZAs), confirmed by the significant statistical validations obtained from two to three ZAs, and the greatest reductions in P2O5 from these two or three ZAs respectively (74 or 40 kg $P2O5$). Thus, this thesis has demonstrated the importance of taking into account the VS of P for specific P management purposes, via the ATV and the delimitation of ZAs with a view to reducing P losses in field crop soils in a precision agriculture context.

Recommendations and outlook

P stratification in agricultural soils remains an important agri-environmental issue. In line with the results of this study, a soil sampling strategy focusing on the top five centimetres should be recommended in soils under grassland and SD for sustainable $P2O5$ recommendations, due to the high risk of soil P stratification. The aim is to reduce P losses to the environment from P-rich fertilisation in relation to these agronomic practices.

To our knowledge, this doctoral thesis represents the first study of the spatial variability of P in grassland soils in the Province of Quebec. However, this study was carried out on a smaller field (2.5 ha). Consequently, additional studies in larger grasslands would be necessary to better assess the spatial structure of P in order to better control the spatial variability of P .

The geostatistical models traditionally used (spherical, exponential) to determine soil P indices are not adapted to soils under SD. Few studies have been carried out to develop geostatistical P prediction models specific to SD soils. Future research should be carried out to develop improved geostatistical models, including cyclic models for better prediction of the spatial structure of P indicators adapted to soils under SD. The aim is to obtain more reliable maps predicting the spatial distribution of P indicators in soils under SD for a better agronomic strategy for controlling P variability in PA.

To our knowledge, this study represents the first experiment on the delimitation of ZAs using the soil ECa-P relationship to control the spatial variability of soil P in a no-till field (P stratified soil) in the Province of Quebec. However, the work was carried out on a single field under SD. Additional studies in other fields would be required to assess the long-term impacts of contrasting tillage practices (CT vs. SD) and soil moisture content on the ECa-P relationship in Quebec soils. The aim would be to improve prediction of the spatial distribution of the agri-environmental P index $(PZAI)_{M3}$ using ECa for sustainable P2O5 recommendations in PA.

The physico-chemical properties of soils, including P, yields and climatic conditions may vary each year as a result of climatic variations, especially in the current context of increasing climate change. Thus, the optimal number of ZAs may change over time and could depend on weather conditions and the crop selected. Few studies have been carried out on the temporal variability of soil P in relation to the temporal variability of crop yields aimed at precisely delimiting temporal ZAs. Future studies would be needed to characterise the temporal variability of soil P in relation to the temporal variability of crop yields in order to precisely delimit temporal ZAs (on a 3-5 year basis, for example). The aim would be to assess the temporary stability of P-based ZAs and the P-yield relationship in a spatio-temporal context in PA. Thus, the choice of the future number of ZAs should be based on a series of temporal data.

The results in Chapter 3 showed that the smallest spatial range values were for P indicators in the top five centimetres of soil under SD in this study. These geostatistical measurements impart on the optimal sampling grid for P and other soil physicochemical properties for sample collection. In addition, the sampling density has an impact on the spatial structure of P. Furthermore, the ZAs delimitation strategy based on the measurement of $(P/Al)_{M3}$ data was the best method of controlling the spatial variability of P, confirmed by the greatest reduction (1-2 ZAs) in the spatial variability of $(P/Al)_{M3}$ (45%), the significant validation tests (2-3 ZAs), and the reduction in the respective inputs of P O_{25} for two and three ZAs (74 to 40 kg P O_{25}). It would therefore be necessary to develop other agronomic strategies for controlling the spatial variability of P, based on the concept of economic optimum sampling density (EOSD) for P, for a better assessment of the spatial structure of soil P in the context of the specific management of PA fields. Few studies have been carried out on the WDEL of P in agricultural environments. Thus, the WDET could be an important measure that deserves attention in future research.

This thesis has demonstrated the importance of characterising the spatial variability of P under different cropping systems with a view to reducing P losses. Few studies have been carried out on the application of optical sensor and gamma ray (γ) spectroscopic methods for characterising the spatial variability of P in 3D. Consequently, it would be necessary to develop other methods for characterising the spatial variability of P in 3D, using spectroscopic techniques with optical sensors and gamma rays. Spectroscopic methods use optical sensors covering wavelengths such as ultraviolet (100-400 nm), visible (400-750 nm), near infrared (750-2500 nm) and mid-infrared (2500-25000 nm). They require little preparation, no logistics and no chemicals to analyse soil samples. Spectroscopic techniques using optical sensors are based on calibration equations between the optical signal and the physicochemical parameter being measured (e.g. ChrysaLabs probes), hence the importance of collecting samples in order to obtain a database. Calibration strategies for optical sensors vary from simple linear regression to multivariate methods. Gamma (γ) spectroscopy measures the intensity distribution of γ radiation as a function of the energy of each photon. It is an important tool for characterising soil properties using point data measurements (e.g. SoilOptix), hence the establishment of sampling campaigns to obtain a database. As yet little used, spectroscopic methods (optical sensors and y-rays) could be studied because of their very promising future for characterising the spatial variability of P indices, particularly in P-stratified soils, with a view to reducing P losses to the environment.

General bibliography

Abdi, D., Cade-Menun, B.J., Ziadi, N., Parent, L.E. (2014). "Long-term impact of tillage practices and phosphorus fertilization on soil phosphorus forms as determined by 31 P nuclear magnetic resonance spectroscopy". Journal of Environmental Quality **43,** 1431-1441.

Adamchuk, V., Allred, B., Doolittle, J., Grote, K., Rossel, R., Ditzler, C., West, L. (2015). Tools for proximal soil sensing. Soil Survey Staff, Ditzler, C., West, L.(Eds.), Soil Survey Manual. Natural Resources Conservation Service. US Department of Agriculture Handbook, **18.**

Allaire, S. E., Van Bochove, E., Denault, J.-T., Dadfar, H., Thériault, G., Charles, A., De Jong, R. (2011). "Preferential pathways of phosphorus movement from agricultural land to water bodies in the Canadian Great Lakes basin: A predictive tool". Canadian Journal of Soil Science **91**(3):361-374.

Ameer, S., Cheema, M.J.M., Khan, M.A., Amjad, M., Noor, M., Wei, L. (2022). "Delineation of nutrient management zones for precise fertilizer management in wheat crop using geo-statistical techniques". Soil Use and Management, **38**(3), 1430-1445.

Beauchemin, S., Hesterberg, D., Chou, J., Beauchemin, M., Simard, R.R., Sayers, D.E. (2003). "Speciation of phosphorus in phosphorus-enriched agricultural soils using X-ray absorption near-edge structure spectroscopy and chemical fractionation". Journal of Environmental Quality **32,** 1809-1819.

Becker, S.M., Franz, T.E., Abimbola, O., Steele, D.D., Flores, J.P., Jia, X., Scherer, T.F., Rudnick, D.R., Neale, C.M. (2022). "Feasibility assessment on use of proximal geophysical sensors to support precision management". Vadose Zone Journal **21**(6), e20228.

Bogunovic, I., Mesic, M., Zgorelee, Z., Jurisic, A., Bilandzija, D. (2014). "Spatial variation of soil nutrients on sandy-loamy soil". Soil and Tillage Research **144**: 174-183.

Brevik, E.C., Fenton, T.E., Lazari, A. (2006). "Soil electrical conductivity as a function of soil water content and implications for soil mapping". Precision Agriculture **7,** 393-404.

Cade-Menun, B., Carter, M., James, D., Liu, C. (2010). "Phosphorus forms and chemistry in the soil profile under long-term conservation tillage: A phosphorus-31 nuclear magnetic resonance study". Journal of Environmental Quality **39**(5): 1647-1656.

Cambouris, A., Messiga, A., Ziadi, N., Perron, I., Morel, C. (2017). "Decimetric-scale two-dimensional distribution of soil phosphorus after 20 years of tillage management and maintenance phosphorus fertilization". Soil Science Society of America Journal **81**(6): 1606-1614.

Cambouris, A., Nolin, M., Zebarth, B., Laverdière, M. (2006). "Soil management zones delineated by electrical conductivity to characterize spatial and temporal variations in potato yield and in soil properties". American Journal of Potato Research **83**(5):381-395.

Corwin, D., Plant R. (2005). "Applications of apparent soil electrical conductivity in precision agriculture". Computers and Electronics in Agriculture **46**(1):1-10.

Dalchiavon, F.C., Rodrigues, A.R., De Lima, E., Lovera, L.H., Montanari, R. (2017). "Spatial variability of chemical attributes of soil cropped with soybean under no-tillage". Revista de Ciências Agroveterinárias **16,** 144-154.

Diacono, M., Castrignanò, A., Troccoli, A., De Benedetto, D., Basso, B., Rubino, P. (2012). "Spatial and temporal variability of wheat grain yield and quality in a Mediterranean environment: A multivariate geostatistical approach". Field Crops Research **131,** 49-62.

Eriksson, A.K., Hesterberg, D., Klysubun, W., Gustafsson, J.P. (2016). "Phosphorus dynamics in Swedish agricultural soils as influenced by fertilization and mineralogical properties: Insights gained from batch experiments and XANES spectroscopy". Science of the Total Environment **566,** 1410-1419.

Fu, W., Zhao, K., Jiang, P., Ye, Z., Tunney, H., Zhang, C. (2013a). "Field-scale variability of soil test phosphorus and other nutrients in grasslands under long-term agricultural managements". Soil Research **51**(6):503-512.

Fu, W., Zhao, K., Tunney, H., Zhang, C. (2013b). "Using GIS and geostatistics to optimize soil phosphorus and magnesium sampling in temperate grassland". Soil Science **178,** 240-247.

Grandt, S., Ketterings, Q.M., Lembo Jr, A.J., Vermeylen, F. (2010). "In-field variability of soil test phosphorus and implications for agronomic and environmental phosphorus management". Soil Science Society of America

Journal **74**, 1800-1807.

Grant, R., Laubel, A., Kronvang, B., Andersen, H., Svendsen, L., Fuglsang, A. (1996). "Loss of dissolved and particulate phosphorus from arable catchments by subsurface drainage". Water Research **30**(11):2633-2642.

Guedes Filho, O., Vieira, S.R., Chiba, M.K., Nagumo, C.H., Dechen, S.C.F. (2010). "Spatial and temporal variability of crop yield and some Rhodic Hapludox properties under no-tillage". Revista Brasileira de Ciência do Solo **34**, 1-14.

Heiniger, R., McBride, R., Clay, D. E. (2003). "Using soil electrical conductivity to improve nutrient management". Agronomy Journal **95**(3): 508-519.

Jamieson, A., Madramootoo, C., Enright, P. (2003). "Phosphorus losses in surface and subsurface runoff from a snowmelt event on an agricultural field in Quebec". Canadian Biosystems Engineering **45**:1-1.

Jia, S.; Zhou, D., Xu, D. (2011). "The temporal and spatial variability of soil properties in an agricultural system as affected by farming practices in the past 25 years". Journal of Food, Agriculture and Environment **9**: 669-676.

Jiang, P., Thelen, K. D. (2004). "Effect of soil and topographic properties on crop yield in a North-Central cornsoybean cropping system". Agronomy journal **96**: 252-258.

Jordan, C., McGuckin, S., Smith, R. (2000) "Increased predicted losses of phosphorus to surface waters from soils with high Olsen-P concentrations". Soil Use and Management **16**(1):27-35.

Jung, W., Kitchen, N., Sudduth, K.A., Kremer, R., Motavalli, P. (2005). "Relationship of apparent soil electrical conductivity to claypan soil properties". Soil Science Society of America Journal **69**(3): 883-892.

Kitchen, N., Westfall, D., Havlin, J. (1990). "Soil sampling under no-till banded phosphorus". Soil Science Society of America Journal **54**:1661-1665.

Kuo, S., Huang, B., Bembenek, R. (2005). "Effects of long-term phosphorus fertilization and winter cover cropping on soil phosphorus transformations in less weathered soils". Biology and Fertility of Soils **41**: 116-123.

Limousin, G., Tessier, D. (2007). "Effects of no-tillage on chemical gradients and topsoil acidification". Soil and Tillage Research **92**(1-2): 167-174.

Mallarino, A. (1996). "Spatial variability patterns of phosphorus and potassium in no-tilled soils for two sampling scales". Soil Science Society of America Journal **60**(5): 1473-1481.

Malvezi, K.E.D., Júnior, L.Z., Guimarães, E.C., Vieira, S.R., Pereira, N. (2019). "Soil chemical attributes variability under tillage and no-tillage in a long-term experiment in southern Brazil". Bioscience Journal **35**(2): 467-476.

Mccormick, S., Jordan, C., Bailey, J. (2009). "Within and between-field spatial variation in soil phosphorus in permanent grassland". Precision Agriculture **10**: 262-276.

Messiga, A.J., Ziadi, N., Morel, C., Grant, C., Tremblay, G., Lamarre, G., Parent, L.-E. (2012). "Long term impact of tillage practices and biennial P and N fertilization on maize and soybean yields and soil P status". Field Crops Research **133**: 10-22.

Messiga, A.J., Ziadi, N., Morel, C., Parent, L.-E. (2010). "Soil phosphorus availability in no-till versus conventional tillage following freezing and thawing cycles". Canadian Journal of Soil Science **90**: 419428.

Messiga, A. (2010). Phosphorus transfers in field crop soils. Doctoral thesis in Soils and Environment, Québec, Université Laval. 217 pages.

Motavalli, P., Hammer, R., Bardhan, S. (2015). "Apparent soil electrical conductivity used to determine soil phosphorus variability in poultry litter-amended pastures". American Journal of Experimental Agriculture **3**(1):287-309.

Nolin, M. and Caillier, M. (1992b). "La variabilité des sols. II-Quantification et amplitude". Agrosol **5**(1):21-32.

Nolin, M., Caillier, M., and Wang, C. (1991). "Soil variability and sampling strategy in detailed soil surveys of the Montreal Plain". Canadian Journal of Soil Science **71**: 439-451.

Nolin, M., Gagnon, B., Leclerc, M., Cambouris, A., Bélanger, G., Simard, R. (2003). Influence of pedodiversity and past land uses on the within-field spatial variability of selected soil and forage quality indicators. In: Proceedings of the 6th International Conference on Precision Agriculture and Other Precision Resources Management, Minneapolis, MN, USA, 14-17 July, 2003. American Society of Agronomy, p 261-277.

Nolin, M.C., Simard, R., Cambouris, A., Beauchemin, S. (1999). Specific Variability of Phosphorus Status and Sorption Characteristics in Clay Soils of the St-Lawrence Lowlands (Quebec). In: Proceedings of the Fourth International Conference on Precision Agriculture, Madison, WI, USA. American Society of Agronomy, Crop

Science Society of America, Soil Science Society of America, p 395-406.

Nyéki, A., Daróczy, B., Kerepesi, C., Neményi, M., Kovács, A.J. (2022). "Spatial Variability of Soil Properties and Its Effect on Maize Yields within Field-A Case Study in Hungary". Agronomy MDPI: **12**(2), 395.

Nze Memiaghe, J.D., Cambouris, A.N., Ziadi, N., Karam, A. (2022). Impacts of Interpolating Methods on Soil Agri-environmental Phosphorus Maps Under Corn Production. In: 15th International Conference on Precision Agriculture, Minneapolis, Minnesota, USA. International Society of Precision Agriculture.

Omonode, R.A., Vyn, T.J. (2006). "Spatial dependence and relationships of electrical conductivity to soil organic matter, phosphorus, and potassium". Soil Science: **171** (3):223-238.

Patoine, M., Hébert, S., Simoneau, M., D'auteuil-Potvin, F. (2017). Loads of phosphorus, nitrogen and suspended solids at the mouths of Quebec rivers, 2009 to 2012. Ministère du Développement durable, de l'Environnement et de la Lutte contre les changements climatiques, Direction générale du suivi de l'état de l'environnement.

Peralta, N.R., Costa, J.L. (2013). "Delineation of management zones with soil apparent electrical conductivity to improve nutrient management". Computers and Electronics in Agriculture **99**: 218-226.

Perron, I., Cambouris, A.N., Chokmani, K., Vargas Gutierrez, M.F., Zebarth, B.J., Moreau, G., Biswas, A., Adamchuk, V. (2018). "Delineating soil management zones using a proximal soil sensing system in two commercial potato fields in New Brunswick, Canada". Canadian Journal of Soil Science **98**: 724-737.

Puustinen, M., Koskiaho, J., Peltonen, K. (2005). "Influence of cultivation methods on suspended solids and phosphorus concentrations in surface runoff on clayey sloped fields in boreal climate". Agriculture, Ecosystems and Environment **105**: 565-579.

Quenum, M., Nolin, M., Bernier, M. (2012). "Numerical mapping of the maximum phosphorus sorption capacity of soils at the agricultural plot scale using auxiliary variables". Canadian Journal of Soil Science **92**: 733-750.

Radocaj, D., Jug, I., Vukadinovic, V., Jurisic, M., Gasparovic, M. (2021). "The Effect of soil sampling density and spatial autocorrelation on interpolation accuracy of chemical soil properties in arable cropland". Agronomy MDPI **11** (12): 2430.

Rhoades, J., Manteghi, N., Shouse, P., Alves, W.J. (1989). "Soil electrical conductivity and soil salinity: New formulations and calibrations". Soil Science Society of America Journal **53** (2): 433-439.

Rodrigues, M.S., Corá, J.E. (2015). "Management zones using fuzzy clustering based on spatial-temporal variability of soil and corn yield". Engenharia Agrícola **35**: 470-483.

Roger, A., Libohova, Z., Rossier, N., Joost, S., Maltas, A., Frossard, E., Sinaj, S. (2014). "Spatial variability of soil phosphorus in the Fribourg canton, Switzerland". Geoderma **217**:26-36.

Saifuzzaman, M., Adamchuk, V., Biswas, A., Rabe, N. (2021). "High-density proximal soil sensing data and topographic derivatives to characterise field variability". Biosystems Engineering **211**: 19-34.

Shi, Z., Wang, K., Bailey, J., Jordan, C., Higgins, A. (2000) "Sampling strategies for mapping soil phosphorus and soil potassium distributions in cool temperate grassland". Precision Agriculture **2**: 347-357.

Singh, G., Williard, K.W., Schoonover, J.E. (2016). "Spatial relationship of apparent soil electrical conductivity with crop yields and soil properties at different topographic positions in a small agricultural watershed". Agronomy MDPI **6** (4): 57.

Soane, B.D., Ball, B.C., Arvidsson, J., Basch, G., Moreno, F., Roger-Estrade, J. (2012). "No-till in northern, western and south-western Europe: A review of problems and opportunities for crop production and the environment". Soil and Tillage Research **118**: 66-87.

Sudduth, K., Kitchen, N., Wiebold, W., Batchelor, W., Bollero, G., Bullock, D., Clay, D., Palm, H., Pierce, F., Schuler, R., Thelen, K. (2005). "Relating apparent electrical conductivity to soil properties across the north-central USA". Computers and Electronics in Agriculture **46**(1-3):263-283.

Sun, W., Huang, B., Qu, M., Tian, K., Yao, L., Fu, M., Yin, L. (2015). "Effect of farming practices on the variability of phosphorus status in intensively managed soils". Pedosphere **25**: 438-449.

Tunney, H. (1990). "A note on a balance sheet approach to estimating the phosphorus fertiliser needs of agriculture". Irish Journal of Agricultural Research **29**(2):149-154.

Tyler, D., Howard, D. (1991). "Soil sampling patterns for assessing no-tillage fertilization techniques". Journal of fertilizer issues.

West, C., Mallarino, A., Wedin, W., Marx, D. (1989). "Spatial variability of soil chemical properties in grazed pastures". Soil Science Society of America Journal **53**(3):784-789.

Whelan, B., McBratney, A. (2000) "The "null hypothesis" of precision agriculture management". Precision Agriculture **2**(3):265-279.

Wilding, L.P., Drees, L. (1983). Spatial variability and pedology. In Developments in Soil Sciences. Elsevier (Vol. 11, pp. 83-116). p. 83-116.

Printed by Books on Demand GmbH, Norderstedt / Germany